Science: A First Course

Science: A First Course

R G Meadows

Hutchinson
London Sydney Auckland Johannesburg

Hutchinson Education

An imprint of Century Hutchinson Ltd
62–65 Chandos Place, London WC2N 4NW

Century Hutchinson Australia Pty Ltd
89–91 Albion Street, Surry Hills,
New South Wales 2010, Australia

Century Hutchinson New Zealand Limited
PO Box 40-086, Glenfield, Auckland 10,
New Zealand

Century Hutchinson South Africa (Pty) Ltd
PO Box 337, Bergvlei, 2012 South Africa

First published in 1989

Set in 10/12pt Times Roman
by Activity Ltd., Salisbury, Wilts

Printed and bound in Great Britain by
Scotprint Ltd., Musselburgh

British Library Cataloguing in Publication Data

Meadows, R.G. (Richard Guy)
Science: a first course.
1. Science
I. Title
500

ISBN 0–09–182361–7

Contents

Preface

This book is written for students taking a first course in science. The contents follow exactly the learning objectives of the first level core unit of science as published by the Business and Technician Education Council (BTEC).

Special emphasis has been placed in providing an easy-to-understand text backed at all stages by fully worked examples. Each chapter also contains a short test and problems (with answers given) to aid revision and assessment.

The writing of this book in parallel with *Mathematics: A First Course* has been an interesting and exciting project and I sincerely hope all students will find both books helpful in strengthening their understanding and application to practical problem solving.

Richard Meadows
April 1989

The International System of units: SI units

In science and engineering and indeed in most other practical areas SI units are now employed. SI is the abbreviation for the International System of units. This system has been adopted by the International Standards Organization (ISO) and by the International Electrotechnical Commission (IEC). We will employ SI units throughout this book.

The International System of units comprises SI units and SI prefixes, and their associated symbols have been internationally agreed. SI units are of three kinds: base, supplementary and derived.

A The seven SI base units and two SI supplementary units

The SI base units for each of the seven physical quantities:

length, mass, time, electric current, temperature, luminous intensity, amount of substance

Table A The seven SI base units and the two SI supplementary units

Quantity	*Name of unit*	*Unit abbreviation*
length	metre	m
mass	kilogram	kg
time	second	s
electric current	ampere	A
thermodynamic temperature	kelvin	K
luminous intensity	candela	cd
amount of substance	mole	mol
plane angle	radian	rad
solid angle	steradian	sr

together with the two SI supplementary units:

radian and *steradian*

used to quantify angular measure, are listed in Table A.

B SI derived units

Units for all other physical quantities are derived from the SI base units. Many of the derived units have their own special names. Table B lists the SI derived units for most of the important physical quantities that you will be meeting in your studies.

C Units of sufficient importance retained for general use

In drawing up the SI system of units it was realized that a number of units were of sufficient practical importance and so widely used that they should be retained and their use still permitted. Table C lists these units.

D SI prefixes

SI prefixes are used to form decimal multiples and submultiples of the SI units. They enable us, together with the unit symbol, to write down the numerical values of physical quantities in a concise and clear way. The symbols, names and meanings of some of the most commonly used SI prefixes are listed in Table D.

The preferred prefixes are in multiples of 3, i.e. ±3, ±6, ±9, ±12. In science and engineering, physical quantity magnitudes are normally specified using the preferred prefixes. When numbers are expressed using $10^{\pm 3}$, $10^{\pm 6}$ ..., etc. multipliers and where the number in front of these multipliers lies in the range 0.1 to 1000, we say the number is expressed *in preferred standard form.*

Table B SI derived units for important physical quantities

Quantity	*Name of unit*	*Unit abbreviation*
area	square metre	m^2
volume	cubic metre	m^3
speed, velocity	metre per second	m/s
angular velocity	radian per second	rad/s
acceleration	metre per second squared	m/s^2
angular acceleration	radian per second squared	rad/s^2
density	kilogram per cubic metre	kg/m^3
frequency	hertz	Hz
force	newton	N
pressure	pascal, newton per square metre	Pa, N/m^2
torque, moment	newton metre	Nm
energy, work	joule	J
power	watt	W
electric charge	coulomb	C
electric potential	volt	V
electric field strength	volt per metre	V/m
magnetic flux	weber	Wb
magnetic flux density	tesla	T
magnetic field strength	ampere per metre	A/m
resistance	ohm	Ω
conductance	siemens	S
capacitance	farad	F
inductance	henry	H
permittivity	farad per metre	F/m
permeability	henry per metre	H/m
specific heat capacity	joule per kilogram kelvin	J/(kgK), $Jkg^{-1}K^{-1}$
specific latent heat capacity	joule per kilogram	J/kg
internal energy, enthalpy	joule	J
specific enthalpy	joule per kilogram	J/kg
specific volume	cubic metre per kilogram	m^3/kg
thermal conductivity	watt per metre kelvin	W/(mK), $Wm^{-1}K^{-1}$

Table C Units of practical importance retained for general use

Name	*Unit symbol*	*Value in SI units*
minute	min	1 min = 60 s
hour	h	1 h = 60 min = 3600 s
degree	°	$1° = (\pi/180)$ rad
litre	l	$1\,l = 10^{-3}\,m^3$
tonne	t	$1\,t = 10^3\,kg$
nautical mile	1 nautical mile = 1852 m	
knot	1 knot = 1 nautical mile per hour	
degree Celsius	°C	$T°C = (T + 273)$ K
standard atmosphere	atm	$1\,atm = 101\,325\,N/m^2$
bar	bar	$1\,bar = 10^5\,N/m^2$

Table D SI prefixes

Name	*Symbol*	*Meaning* (i.e. factor by which unit is multiplied)
tera	T	$\times 10^{12}$ (a million million times)
giga	G	$\times 10^9$ (a thousand million times)
mega	M	$\times 10^6$ (a million times)
kilo	k	$\times 10^3$ (a thousand times)
hecto	h	$\times 10^2$ (a hundred times)
deca	da	$\times 10^1$ (ten times)
deci	d	$\times 10^{-1}$ (a tenth)
centi	c	$\times 10^{-2}$ (a hundredth)
milli	m	$\times 10^{-3}$ (a thousandth)
micro	μ	$\times 10^{-6}$ (a millionth)
nano	n	$\times 10^{-9}$ (a thousand millionth)
pico	p	$\times 10^{-12}$ (a million millionth)

Examples

1 A resistor of value 470 000 ohms can be expressed as

$$470\,000 = 4.7 \times 10^5\,\Omega \text{ in standard form}$$
$$= 470 \times 10^3$$
$$= 470\,k\Omega \text{ in preferred standard form}$$

Note: the prefix k means $\times 10^3$ or $\times 1000$; the symbol Ω is the SI symbol for ohms.

2 $T = 10\,\mu s$ means $T = 10 \times 10^{-6}$ seconds; the prefix μ means $\times 10^{-6}$ or one millionth; s is the SI symbol for seconds.

3 $f = 50$ MHz means $f = 50 \times 10^6$ hertz; the prefix M means $\times 10^6$ (a million times); Hz is the SI symbol for hertz, the unit of frequency.

4 $0.039\,m = 39 \times 10^{-3}\,m$ (m = abbreviation for metre)
$= 39$ mm (mm = millimetres, prefix m denoting 10^{-3})

5 $E = 210\,GN/m^2$ (Young's modulus for steel) means $E = 210 \times 10^9\,N/m^2$; the prefix G means $\times 10^9$; N/m^2 stands for newtons per square metre.

Part One: Oxidation

1 Oxidation

General learning objectives: to establish, through experiments, the basic chemical processes involved in burning and rusting as examples of chemical reactions (interactions between substances which result in a rearrangement of their atoms) and to apply this knowledge to a variety of practical situations.

1.1 The composition of air

Air is a mixture of many gases but the main constituents by far are nitrogen and oxygen. The ratio by volume is approximately a little less than four-fifths nitrogen and slightly greater than one-fifth oxygen.

Air contains also a very small fraction by volume of argon, neon, helium, krypton, xenon (known as the rare gases) and also tiny amounts of carbon dioxide, methane, nitrous oxide, hydrogen and ozone. Obviously, unless air is deliberately dried by some absorbing chemical, it will also contain water vapour. As a result of industrial pollution many other gases and solid particles may additionally be present.

Table 1.1 summarizes the percentage constituents by volume of dry 'clean' air.

Examples

1 The density of air (mass in kilograms per cubic metre) is $1.2\,kg/m^3$ at 20°C. Estimate for a room of dimensions 10 m by 5 m and height 3 m the mass of air present at 20°C. Estimate also the effective volume of oxygen present.

Solution

$$\text{Volume of room} = 10 \times 5 \times 3 = 150\,m^3$$

so mass of air in the room,

$$\text{mass} = \text{volume} \times \text{density} = 150 \times 1.2 = 180\,kg \quad \textit{Ans}$$

The percentage of oxygen by volume is approximately 21 per cent (see Table 1.1), so the effective volume of air in the room,

$$\text{Volume of oxygen} = 21 \text{ per cent of } 150\,m^3 = 0.21 \times 150 = 31.5\,m^3$$

Ans

2 The cylinder shown in Figure 1.1(a) contains air at atmospheric pressure. A chemical which is a good absorber of oxygen is now introduced into the cylinder. State what you think may happen assuming the plunger is held at

Table 1.1 Constituents by volume of dry air (the earth's atmosphere excluding water vapour)

Gas	*Chemical symbol*	*Percentage volume*
Nitrogen	N_2	78.09
Oxygen	O_2	20.95
Argon	Ar	0.93
Carbon dioxide	CO_2	0.03
Neon	Ne	0.0018
Helium	He	0.0005
Methane	CH_4	0.00015
Krypton	Kr	0.00011
Nitrous oxide	N_2O	0.00005
Hydrogen	H_2	0.00005
Ozone	O_3	0.00004
Xenon	Xe	0.00001

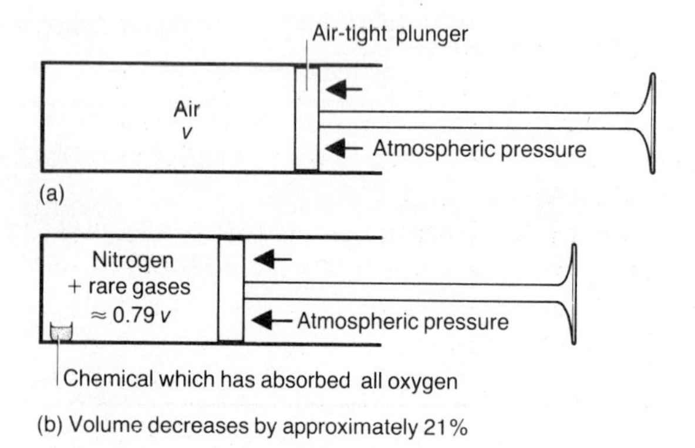

Figure 1.1

atmospheric pressure and that any temperature changes that may have occurred in the absorption process have settled down.

Solution

Since the chemical absorbs all the oxygen in the air the cylinder will now contain only nitrogen plus a small fraction of rare gases, etc. The plunger will move in and reduce the volume by about 21 per cent (the effective volume occupied by the original oxygen), as shown in Figure 1.1(b).

1.2 Mass gain of metals when heated in air

When most metals are heated and burn in air they undergo a chemical change and the resulting substance formed is normally very different in its look and properties from the original metal.

For example, when copper foil is heated in a Bunsen flame it becomes coated with a blackish tarnish. When lead shot is heated in an iron spoon it becomes coated with a yellow powder. If zinc is heated strongly it becomes coated with a yellow powder, which changes its colour to white on cooling. A piece of magnesium, when heated in a Bunsen flame, gives off a brilliant white flame and undergoes a chemical change from metallic form to a white powder.

What has occurred in these processes? In each of the above examples, the metal when heated takes oxygen from the air and combines with the oxygen to form a compound. The mass of the compound so formed is greater than the original mass of the metal.

The fact that an increase in mass takes place can be verified experimentally using the simple apparatus shown in Figure 1.2. A crucible together with its lid and containing some copper foil or magnesium ribbon is first weighed. The weight is recorded. The crucible and contents are then heated and the metal begins to burn. The lid is lifted a few times during heating to allow more air (i.e. the oxygen in the air) to combine with the metal. Care should be taken to ensure that as little 'smoke' as possible escapes when the lid is lifted. When all or at least most of the metal is burned the crucible containing its contents are re-weighed. It will be observed that an increase of mass has taken place.

The increase in mass results because the metal has combined with oxygen in the air to form a compound of metal and oxygen. The actual increase is the mass of oxygen which has reacted with the metal.

Note, however, that although the experiment, if carefully performed, shows a mass increase, it does not provide us with conclusive proof that the gas taken from the air is oxygen. If the metal had been heated in a closed atmosphere of air, it could be observed that burning ceases when approximately one-fifth of the volume of air is used up. This, again, does not provide us with absolute proof, but does strongly suggest that the metal has combined with the 21 per cent of the oxygen contained in the air.

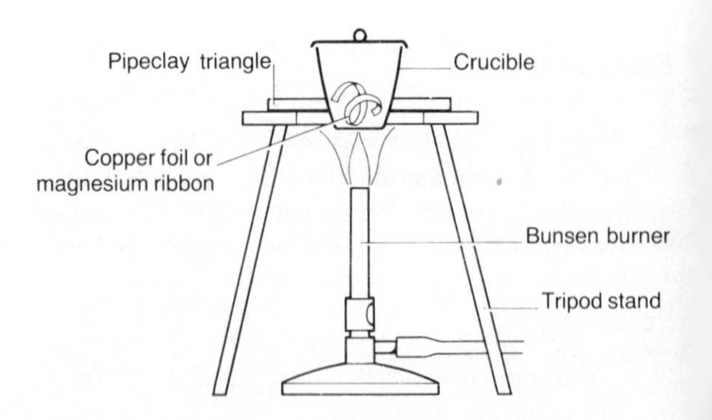

Figure 1.2 Apparatus used to demonstrate the gain of mass when a metal is heated in air

1.3 Analysis of oxides

Many metals, but by no means all, 'burn' in air. Many other elements, such as carbon and sulphur, also burn in air. Some substances require quite vigorous heating before burning commences although some elements, for example sodium, combine rapidly with the oxygen in the air even at room temperature.

The burning process, the reaction with the oxygen in the air is an example of a chemical reaction. Chemical reactions are said to occur when substances interact with each other and the interaction results in a permanent rearrangement of the atoms of those substances. The substances combine to form a *chemical compound*.

When elements, such as carbon, copper, magnesium, lead, sulphur, and phosphorus, burn in air, they combine chemically with the oxygen present in air to form a *compound*. The chemical reaction which takes place is known as oxidation and the resulting compound formed is known as an *oxide*.

Once oxidation commences, it is often self-sustaining. For example, once carbon, say in the form of coke or coal, starts burning it will continue to do so while oxygen is available, giving out heat in the process.

Substances burning in air, or in general undergoing oxidation, combine with oxygen in definite proportions by mass. Substances do not normally combine with nitrogen or any other of the constituent gases in air.

Oxides – the compounds formed by the chemical combination of an element with oxygen – may be in the form of a gas, a liquid or a solid. For example, when carbon burns in air it forms the compound of carbon dioxide and in some cases also carbon monoxide. Both these oxides are gases. When hydrogen combines with oxygen – care, the oxidization process is explosive – it forms water. Water (H_2O) is the oxide of hydrogen. Most metal oxides are solids. Not all metals form oxides when heated in air. Gold, silver, platinum, and mercury are examples of the *noble metals* which are not oxidized when heated in air.

Two examples of an oxidation reaction and an analysis of the oxides formed are considered below:

1 Oxidation of carbon

Although literally billions of atoms will be combining in an oxidation process, the respective amounts of the element and oxygen will always be in definite fixed ratios.

When carbon burns in air each atom of carbon combines with two atoms of oxygen to form one molecule of carbon dioxide. This reaction can be represented by the chemical equation

$$C + O_2 \rightarrow CO_2$$

where C is the chemical symbol for carbon and is also used to represent one carbon atom

O is the chemical symbol for oxygen

O_2 is the symbol for the oxygen molecule which consists of two oxygen atoms

CO_2 is the chemical symbol for carbon dioxide

The equation is essentially a neat, shorthand way of describing the chemical reaction taking place in the oxidation process.

The atomic mass of carbon (C) is 12 units and that of oxygen (O) is 16 units, so the equation tells us the proportions by mass in which carbon and oxygen combine to form carbon dioxide, i.e.

12 units of carbon + $(2 \times 16 = 32)$ units of oxygen $\rightarrow 12 + 32 = 44$ units by mass of carbon dioxide

(*Note:* The atomic mass is defined in units equal to one twelfth of the mass of the carbon atom. On this scale hydrogen (H) has an atomic mass of 1 and 1 atomic mass unit equals approximately 1.66×10^{-27} kg.)

It is also of interest to note that when carbon 'burns' to form carbon dioxide, considerable heat is also given out in the process. For every kilogram of carbon consumed, 33×10^6 joules of heat are given out.

2 Oxidation of magnesium

The metal magnesium (chemical symbol Mg and atomic mass 24.305) combines with oxygen atoms on a one-to-one basis. Two magnesium atoms combine with two oxygen atoms (one oxygen molecule, oxygen exists in air only as molecules)

to form magnesium oxide. The chemical equation is

$$2\,Mg + O_2 \rightarrow 2\,MgO$$

Note that we did not write the equation as $Mg + O \rightarrow MgO$. The reason for this is that the smallest entity of oxygen which can have an independent existence is the oxygen molecule, i.e. O_2, not a single oxygen atom O.

Since the atomic mass of magnesium is 24.305 and that of oxygen is 16 the equation also shows the proportions by mass by which the elements combine to form magnesium oxide (MgO):

$$2 \times 24.305 \text{ of Mg} + 2 \times 16 \text{ of O} \rightarrow 2(24.305 + 16) \text{ of MgO}$$

or more simply dividing throughout by 2,

$$24.305 \text{ parts Mg} + 16 \text{ parts O} \rightarrow 40.305 \text{ parts MgO}$$

1.4 Rusting: the effects of oxygen and water on iron

All of us will be very much aware of how iron left out in the open rapidly rusts and the damage that can be caused by rusting. What is rust and what is happening in the rusting process?

Rusting is the process in which iron changes chemically in reacting with oxygen and water to a complex compound of iron oxide with attached water molecules, which we call *rust*. Rust itself has the approximate chemical formula of $2\,Fe_2O_3 . 3\,H_2O$ (Fe is the chemical symbol for iron; H_2O is the chemical formula for water and Fe_2O_3 the chemical formula for ferric oxide). Rust has the chemical description of hydrated ferric oxide, 'hydrated' to indicate that the oxide contains also water molecules.

Rust is formed on the surface of iron and steel (steel is iron containing from 0.1 to 1.5 per cent of

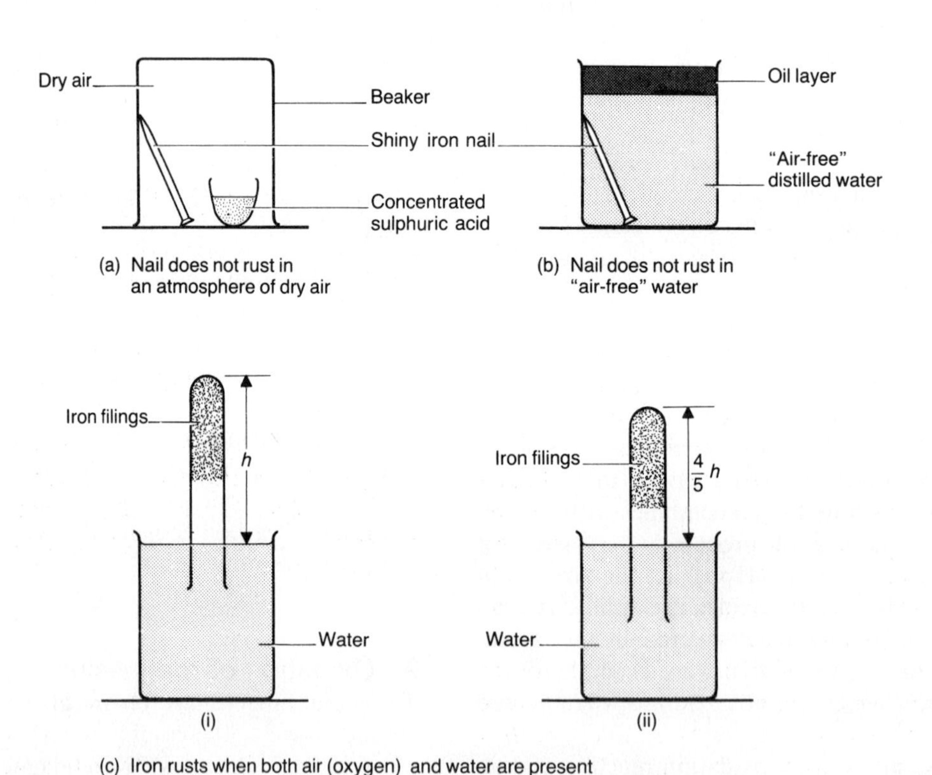

Figure 1.3 Experiment to illustrate that both oxygen and water are necessary to cause rusting

carbon, the carbon being added to produce different properties) when it is exposed to moisture and air. Iron will not rust in perfectly dry air, nor will iron rust in water which is completely free from dissolved oxygen.

The fact that both oxygen and water are involved and indeed required for rusting can be demonstrated in the following experiment.

1 Place an iron or steel nail in a perfectly dry atmosphere of air, as shown for example in Figure 1.3(a). The concentrated sulphuric acid acts as a dehydrating agent, i.e. it absorbs any water vapour in the air. This ensures that no water is present in the air in the beaker and hence no water comes into contact with the nail.

   Observe that no rusting takes place even over a period of a week or more by the fact that the nail remains in its original shiny state. The presence of rust would be indicated by the nail turning a reddish-brown colour.

2 Place some distilled water in a beaker and boil gently to expel any air that may have been absorbed in the water. Immerse a nail in the water and cover the water surface with a layer of oil to ensure no air comes into contact with the water, as shown in Figure 1.3(b). This layer prevents any possibility of air, and therefore oxygen, dissolving in the water. Observe that no rusting occurs over the next few days.

3 Wet the sides of a test tube with water. Place some iron filings in the tube and shake so that they or most of them adhere to the tube sides. Place the tube mouth downwards in a beaker of water, as shown in Figure 1.3(c)(i). Leave for approximately one week.

   After this time it will be observed that the filings will have rusted, and if the water levels in the beaker and tube are equalized (so that the 'air' in the tube is at atmospheric pressure) it will also be noted that the air in the tube is reduced by approximately one-fifth. This indicates that one fifth by volume of the original air has combined with the iron in the rusting process. This one-fifth is oxygen.

Also from observations 1,2 and 3 above, we can conclude that both oxygen and water are necessary for rusting to take place.

1.5 Examples of the damage caused by rusting

It is estimated that the annual costs of metal corrosion amounts to between £50 and £100 per person in an industrialized country. In Britain alone approximately two to three thousand million pounds is lost annually by corrosion. A very high percentage of this is due to the damage caused by the rusting of iron and steel products and structures. It is also estimated that between ten and twenty per cent of the production of iron and steel is used to replace that damaged by rusting.

Figure 1.4 illustrates an example of the damage caused by the rusting of a steel plate with one surface exposed to the atmosphere (i.e. air containing oxygen, water vapour and rain). Outer layers of rust which are formed are only loosely connected to the layers nearest the yet unrusted steel. Although the inner layers of rust adhere most firmly to the steel, the outer layers tend to flake off. As the corroding process produced by rusting continues, the strength of the steel plate is weakened and will eventually fracture if under any pressure, or rust through completely.

Since rusting can and does occur whenever iron and steel is exposed to the atmosphere, examples of the damage caused is almost endless. Add your own examples to this list: rusting of steel and iron components and structures, bridges, masts, bolts, nuts, etc.; rusting of cars, ships, etc. – rusting is normally the major cause of eventually scrapping a car or ending the useful lifetime of a ship; rusting of iron drain-pipes, gutters, window frames, etc.

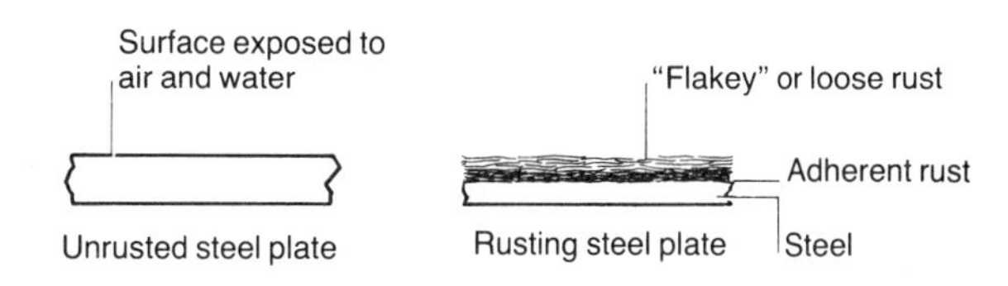

Figure 1.4 An example of the damage done by rusting

1.6 Methods used to prevent rusting

Since the damage caused by rusting is so widespread, needless to say, immense amounts of money and effort are expended to combat and reduce rusting.

The simplest means of preventing rusting is to ensure that air and water, the reactants necessary for rusting, are blocked from the iron and steel. Thus rusting can largely be prevented by painting or greasing exposed iron surfaces. The lifetime of iron or steel mesh fencing can be increased considerably by plastic coating the mesh.

Plating of steel surfaces with another metal is also used extensively. For example, zinc plating, known as galvanizing, produces a protective layer of zinc on steel. The zinc reacts with the oxygen and carbon dioxide in the atmosphere to form a strong adherent coating which is very resistant to rusting.

Another method for protecting iron and steel is to alloy iron with non-corrosive metals such as nickel and chromium. Stainless steel which consists of 18% nickel and 8% chromium has a very high resistance to corrosion.

Test 1

This test may be used as a quick self-assessment to check whether you have absorbed the main facts of the first chapter on **Oxidation** and its objectives. Enter your answer for each question in the appropriate box in the answer block given below.

Enter a tick (√) in the answer block if you consider the statement correct; enter a cross (×) if you consider the statement incorrect, even if part of it might be right.

Qu. 1 The main constituents of the earth's atmosphere (air) and their approximate percentage by volume are:
(a) about 99% nitrogen and oxygen
(b) nitrogen 21%
(c) oxygen 21%
(d) nitrogen 78%

Qu. 2 Air also contains a small fraction by volume of the following gases:
(a) carbon dioxide
(b) the rare gases including argon and neon
(c) hydrogen and ozone

Qu. 3 Magnesium foil is heated strongly in air for several minutes and then allowed to cool. Care is taken so that no material involved in the burning process escapes.
(a) Magnesium loses mass when heated in air since its metallic properties are destroyed.
(b) Magnesium gains mass when heated in air.
(c) Magnesium combines with the oxygen in the air to form magnesium oxide.
(d) The chemical reaction taking place is an example of oxidation.

Qu. 4 The following statements refer to rusting.
(a) Rust is an oxide of iron with additional water molecules in its chemical composition.

Answer block:

Question no.	1				2			3			
	(a)	(b)	(c)	(d)	(a)	(b)	(c)	(a)	(b)	(c)	(d)
Answer											

Question no.	4							5			
	(a)	(b)	(c)	(d)	(e)	(f)	(g)	(a)	(b)	(c)	(d)
Answer											

(b) Rusting occurs rapidly in dry air at high temperatures.
(c) Rusting will occur in 'air-free' distilled water but will take a little longer time.
(d) Both oxygen and water are required for rusting to occur.
(e) Rusting can be prevented by painting the exposed surfaces with an oil-based paint.
(f) Rusting weakens the strength of iron/steel components and structures.
(g) Galvanizing is a process of coating iron or steel with zinc to preserve it from rusting.

Qu. 5 (a) Oxidation is the process where an element (e.g. a metal) combines with oxygen to form an oxide.
(b) Dry air is approximately one-fifth oxygen, four-fifths nitrogen.
(c) Rusting occurs more rapidly if the relative humidity (water vapour content) of the atmosphere is high.
(d) Gold is one of the so-called *noble metals* because it does not oxidize when heated in air.

Problems 1

1 Describe briefly an experiment that you would perform to demonstrate that a metal gains mass when heated in air. Explain the reason for the mass gain.

2 Powdered copper (contained in a crucible plus lid) is heated in air for several minutes and allowed to cool. Care is taken that no fumes from the burning process escape from the crucible. It is observed that the copper turns black.
(a) What elements are involved in the reaction?
(b) Name the reaction and the compound formed.
(c) Explain why the mass of the compound formed is greater than the original mass of copper.

3 The equation defining the oxidation of carbon to carbon dioxide is:

$$C + O_2 \rightarrow CO_2$$

If the atomic mass of carbon (C) and oxygen (O) are respectively 12 and 16, determine the amount of oxygen needed to fully oxidize 100 grams of carbon, and the mass of carbon dioxide formed.

4 A small iron bar is placed in the following three environments:
(a) in dry air, kept dry by a dehydrating agent;
(b) in a beaker containing normal tap water;
(c) close to the steam generated by a boiling kettle.

State in each case whether rusting will occur and give reasons.

5 State the two constituents that are necessary for rusting of iron and steel to take place and describe an experiment to demonstrate that both constituents are necessary.

6 (a) Give three practical examples of the damage done by rusting.
(b) Describe the methods used to prevent rusting.

Part Two: Statics

2 Elasticity and Hooke's law

General learning objectives: to produce graphs from results obtained experimentally to determine the relationship between force and extension for different given materials and subsequently to verify Hooke's law relating to elasticity and to solve practical problems.

2.1 Tensile, compressive and shear forces and stresses

In this chapter we consider mainly the effect of tensile forces acting on materials and in particular we investigate the relationship between tensile force and the extension it produces for different materials. As an introduction to this work, some general definitions and points concerning elasticity and types of forces are first explained.

When materials are subjected to external forces they invariably deform or distort in some manner. Provided that the distortion produced is relatively small most materials return to their original shape when the applied forces are removed. The capacity of materials to return to their original shape when the applied forces are removed is called *elasticity*. If the applied forces are sufficiently large for the deformation to cause breaks in the molecular structure, the material ceases to be entirely elastic and its *elastic limit* is said to have been exceeded.

The application of an applied force to a material causes internal forces to be set up within the material. These internal forces are known as forces of stress. Stress itself is defined in terms of force per unit area produced in the material by the external force, i.e.

$$\text{stress} = \frac{\text{external force applied}}{\text{area over which force acts}}$$

The SI units of force and area are the newton (N) and the square metre (m^2) so the units of stress are newtons per square metre, N/m^2.

There are three main types of force and stress: tensile, compressive and shear. The nature of these is now explained.

2.1.1 Tensile forces and stresses

When the action of the externally applied force tends to pull apart or to stretch the material or component, the latter is said to be under tension and the force producing the tension is known as a tensile force.

Examples of tensile force and components under tension are shown in Figure 2.1. In (a), a bar is clamped at end B and the other end C is subjected to a tensile force. The bar extends somewhat and is in a state of tension. Inside the bar there are internal resisting forces which act in opposition to the applied tensile force. These internal forces are cohesive forces between the grains and within the grains of the bar material and their cumulative effect per unit area is the tensile stress. The magnitude of the stress is determined by dividing the applied force by the cross-sectional area of the bar, i.e.

$$\text{tensile stress} = \frac{\text{tensile force}}{\text{area}} = \frac{F}{A}$$

where F = tensile force in newtons
A = area in square metres

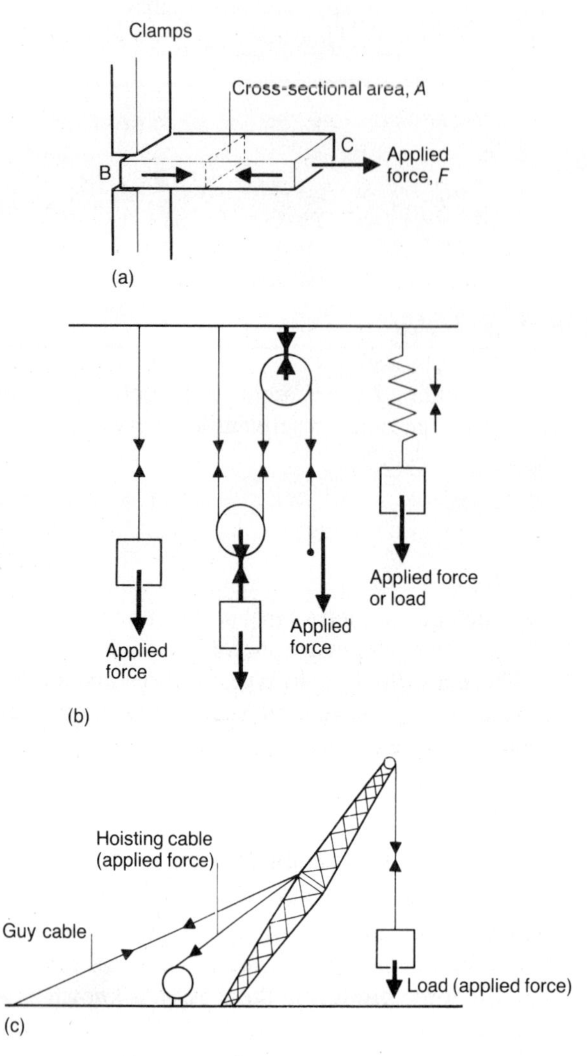

Figure 2.1 Examples of tensile force and stress. (Tensile stress in the individual components is by the two inward arrows → ←)

and the plane of A is at right angles to line of action of F.

Other examples of components undergoing tensile stress are shown in Figure 2.1(b) and (c). The tensile stresses in the ropes and cables in these diagrams are denoted by the two opposing 'inwards' arrows →←.

2.1.2 Compressive forces and stresses

When the action of the externally applied force tends to compress the material or component, the latter is said to be under compression and the force producing the compression is known as a compressive force. Compressive forces create compressive stresses acting in opposition.

Figure 2.2 shows examples of compressive forces and the compressive stresses set up as a result of their application. In the diagrams, compressive stresses are denoted by two opposing 'outward' arrows ← →. Note, however, that in (d) the components AB and BC are actually in a state of tension, while all other parts of the structure are in compression.

2.1.3 Shear forces and stresses

In cases where the applied force is at right angles to the cross-sectional area of the material or component, the force produces either a tensile or compressive stress within the component. A third type of stress, known as shear stress, is caused when the applied force acts parallel to the cross-sectional plane of the component, as shown for example in Figure 2.3(a), where the base of a

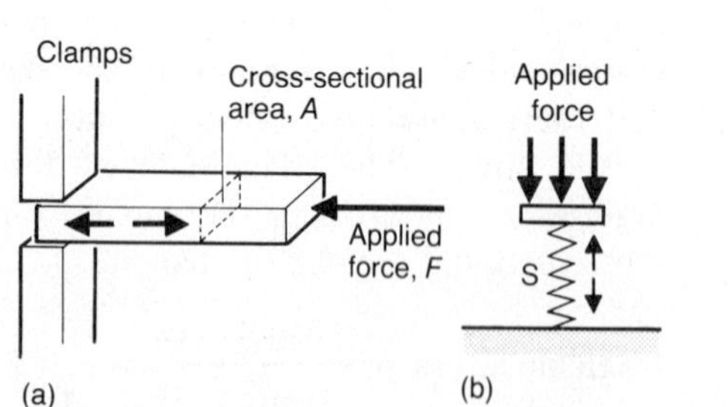

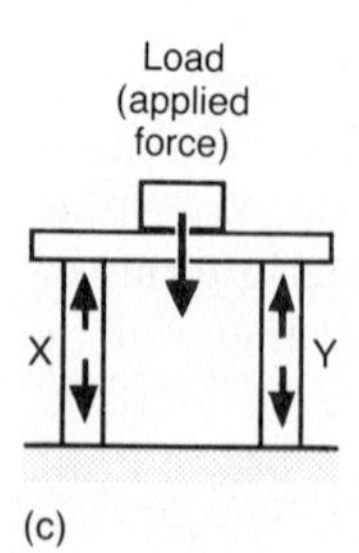

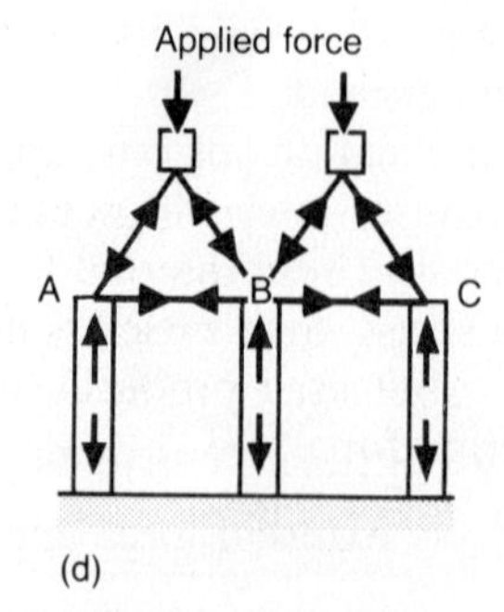

Figure 2.2 Examples of compressive force and stress. (Compressive stress in the individual components is denoted by the two outward arrows ← →)

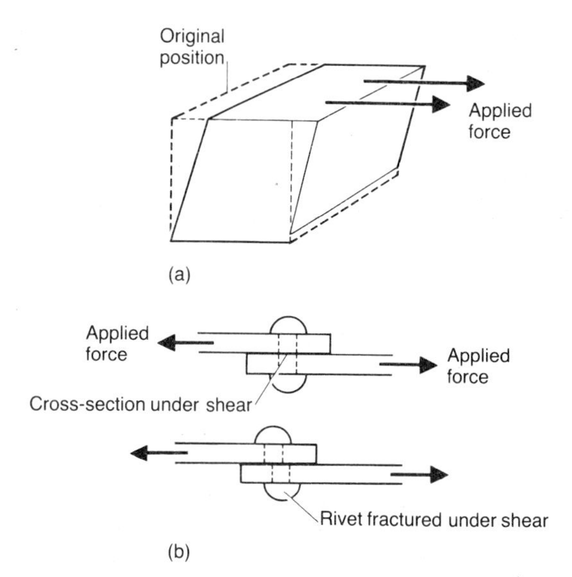

Figure 2.3 Examples of shearing force and stress

cube of material is clamped and a force is applied parallel to the top face of the cube. The cube undergoes a shearing-type of distortion in which the layers of the material tend to slide over adjacent layers.

Figure 2.3(b) shows a practical example of shearing. The purpose of the rivet is to prevent one plate from sliding over the other as well as securing the plates together. Examples of shear forces which actually cause fracture are produced by scissors, hand shears and guillotines.

2.2 Hooke's law relating stress and strain

The relation between the extension produced and the magnitude of the tensile force producing it was investigated by Robert Hooke (1635–1703) and he found that within the elastic limit, *the extension produced was directly proportional to the applied force producing it*, i.e.

$$x \propto F, \quad \text{or} \quad x = kF$$

where x = extension, F = applied force and k is the constant of proportionality.

The above statement forms a simple statement of what has now become known as *Hooke's law*. In order to express Hooke's law in a more general form and also to give a quantitative measure of the deformation in a material or component when subjected to applied forces, we introduce the term strain.

Strain is a measure of the ratio of the deformation produced to an original (unstressed) dimension. The three strains corresponding to tensile, compressive and shearing forces, are defined as:

$$\text{Tensile strain} = \frac{\text{extension}}{\text{original length}} = \frac{x}{L}$$

[Figure 2.4(a)]

$$\text{Compressive strain} = \frac{\text{compression}}{\text{original length}} = \frac{x}{L}$$

[Figure 2.4(b)]

$$\text{Shear strain} = \frac{\text{shear displacement}}{\text{original length}} = \frac{x}{L}$$

[Figure 2.4(c)]

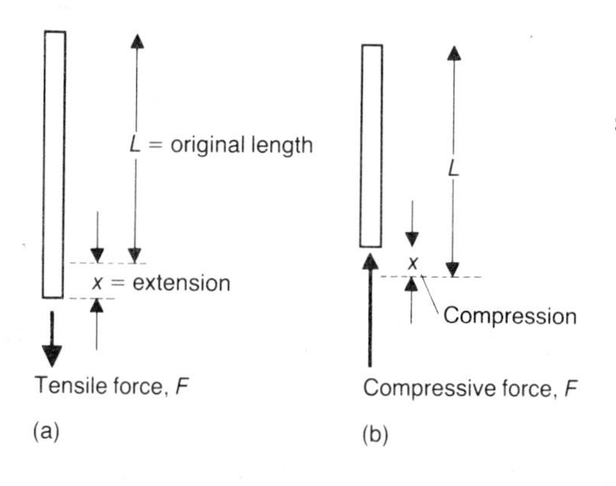

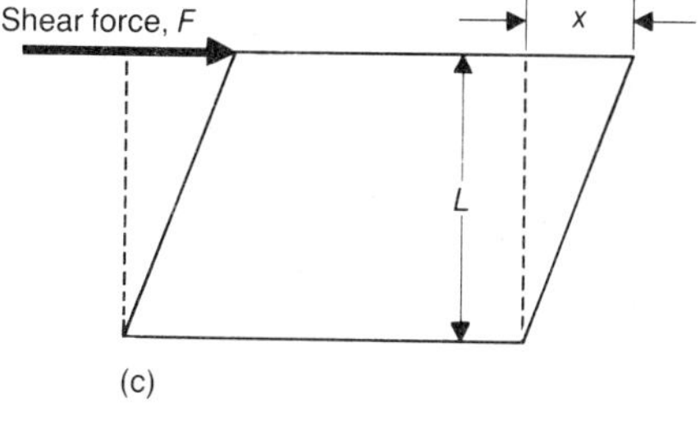

Figure 2.4 Definition of strain

Strain is the ratio of two lengths and therefore has no units.

Thus, remembering that stress is force per unit, we are now in a position to give a general statement of *Hooke's law*:

> *The strain produced in a material is directly proportional to the stress producing it, provided the elastic limit of the material has not been exceeded.*

Hooke's law is closely obeyed by many materials provided the strain is relatively small. For example, for most metals the elastic limit corresponds to strains of the order of 0.1 per cent.

The ratio of stress to strain up to the elastic limit for a material which obeys Hooke's law is a constant. This constant is known as the *modulus of elasticity*, and for tensile and compressive stress the constant is usually known as Young's modulus.

Young's modulus, named after Thomas Young (1773–1829) who investigated the stress–strain relationship for wires and rods, is usually denoted by the letter E:

$$E = \frac{\text{stress}}{\text{strain}}$$

$$= \left(\frac{F}{A}\right) \div \left(\frac{x}{L}\right) = \frac{FL}{xA}\ \text{N/m}^2$$

Table 2.1 Young's modulus of elasticity and tensile strength for some common materials

Material	*Young's modulus* E (N/m^2)	*Tensile strength* (N/m^2)
Aluminium	68 to 72 $\times 10^9$	70 to 150 $\times 10^6$
Copper	96 to 132 $\times 10^9$	120 to 400 $\times 10^6$
Mild steel	190 to 220 $\times 10^9$	420 to 510 $\times 10^6$
Iron (cast)	90 to 130 $\times 10^9$	100 to 230 $\times 10^6$
Iron (wrought)	180 to 195 $\times 10^9$	280 to 450 $\times 10^6$
Concrete	~28 $\times 10^9$	~4 $\times 10^6$
Glass (crown)	~70 $\times 10^9$	30 to 90 $\times 10^6$
Oak (along grain)	~11 $\times 10^9$	60 to 110 $\times 10^6$
Nylon (at 20°C)	0.8 to 3.1 $\times 10^9$	76 to 97 $\times 10^6$

where F = applied force in newtons (N)
A = cross-sectional area of material in square metres (m^2)
x = extension or compression
L = original length

Some typical values for E for some common materials are given in Table 2.1. Values of tensile strength, which is defined as the maximum tensile stress beyond which the material would eventually fracture, are also included. Note steel, as you might expect, is the strongest material.

2.3 Experimental determination of the relationship between force and extension

Figure 2.5 shows a typical experimental set-up which can be used to investigate the relationship between extension and tensile force for a metal wire, nylon rope, length of rubber or even a glass rod. In this particular case we are considering a test on a one-metre length of mild-steel wire of radius 0.564 mm.

$L = 1\,\text{m}$; area $A = \pi r^2$, $r = 0.564 \times 10^{-3}\,\text{m}$
so area $A = \pi r^2 = 3.142 \times (0.564 \times 10^{-3})^2$
$= 1.0 \times 10^{-6}\,\text{m}^2$

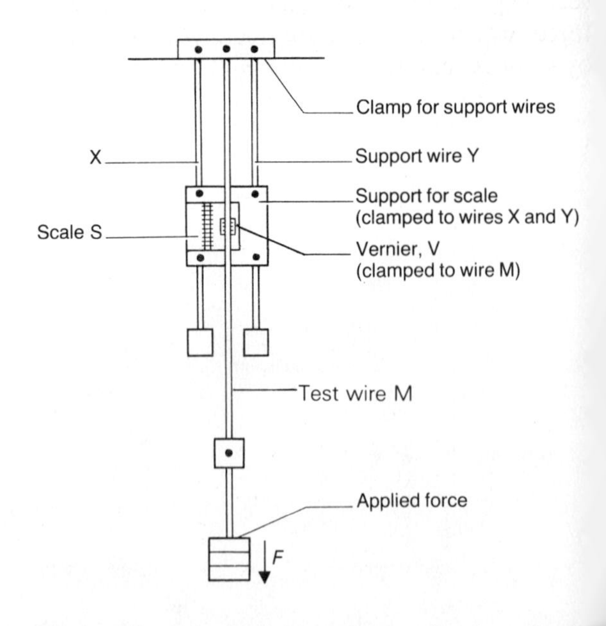

Figure 2.5 Experimental apparatus for investigating extension versus tensile force relationships

A vernier extensometer V is clamped to the test wire M. Wires X and Y are used to carry the scale S. All three wires are clamped to a common solid support. Thus even if the support were to sag when loads were applied to M, the scale would also drop by an equal amount and annul any false extension reading. Vernier V in conjunction with scale S is used to measure the extensions produced as the load (tensile force) is steadily increased.

Note: the load should include the weight of the vernier plus the carriage attached to wire M in addition to the actual weights added. The total applied force (load),

$$F = mg$$

where m = total mass acting on wire, in kilograms (kg);
g = 9.81 m/s^2, the acceleration due to gravity.

The following results were obtained by applying loads of increasing magnitude of 20 N steps. Up to loads of 160 N a check was made to see whether or not the elastic limit of the wire had been exceeded. This was accomplished by removing the load at each step and noting that the wire returned to its unstressed length.

Force, F (N)	0	20	40	60	80
Extension, x (mm)	0	0.10	0.19	0.29	0.37
Force, F (N)	100	120	140	160	180
Extension, x (mm)	0.48	0.56	0.65	0.76	0.88
Force, F (N)	200	220	240	260	
Extension, x (mm)	1.04	1.20	1.43	1.76	

A graph of extension x versus tensile force (load) F is now plotted and this is shown in Figure 2.6. A smooth curve is drawn to 'best-fit' the plotted points. For loads up to approximately 160 N the graph is very closely a straight line and

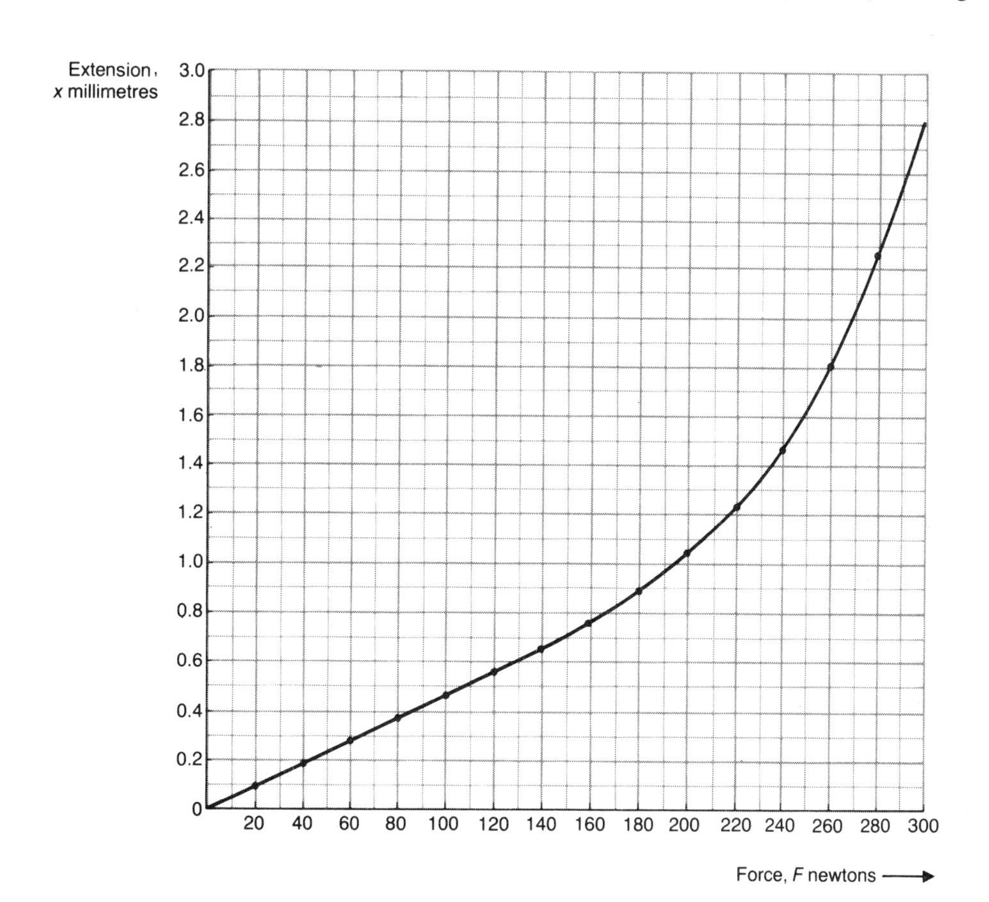

igure 2.6 Plot of experimental results of extension versus tensile force for a 1 m length of steel wire of iameter 1.128 mm

hence we can conclude over this range that extension is directly proportional to applied force and Hooke's law applies to a very good degree, certainly within the experimental accuracy of our measurements. At loads greater than 160 N the graph deviates considerably from a straight line and a given increase in load causes a much greater extension than in the Hooke's law region. If loading is further increased beyond 280 N or so fracture will almost certainly occur.

Using the above results we can also plot stress versus strain over the linear region and deduce the value of Young's modulus of elasticity, E, for the steel of the test specimen:

$$\text{Stress} = E \times \text{strain}$$
$$F/A = E \times (x/L)$$

Thus if stress is plotted (y-axis) against strain (x-axis) E can be determined by measuring the gradient of the resulting straight line. The values for stress (found by dividing the force F by the cross-sectional area A of the wire) and strain (found by dividing the extension x by the original length L) for the range 0 to 160 N (with $A = 1.0 \times 10^{-6}\,\text{m}^2$ and $L = 1$ m), are given below:

Force, F (N)	0	20	40	60	80
Stress, F/A (MN/m^2)	0	20	40	60	80
Strain, x/L ($\times 10^{-3}$)	0	0.1	0.19	0.29	0.37

Force, F (N)	100	120	140	160
Stress, F/A (MN/m^2)	100	120	140	160
Strain, x/L ($\times 10^{-3}$)	0.48	0.56	0.65	0.76

The graph of stress versus strain is plotted in Figure 2.7 and the gradient of the straight line drawn through the points is:

$$\frac{a}{b} = \frac{(149 - 14)\ \text{MN/m}^2}{(0.7 - 0.065) \times 10^{-3}}$$

$$= \frac{135 \times 10^6}{0.635 \times 10^{-3}} = 212.6 \times 10^9\ \text{N/m}^2 \text{ or } 212.6\ \text{GN/m}^2$$

which provides us with the value of Young's modulus of elasticity, i.e.

$$E = \text{gradient of stress–strain curve}$$
$$= 213\ \text{GN/m}^2.$$

Remember:

1 GN (one giga-newton) $= 10^9$ N
1 MN (one mega-newton) $= 10^6$ N

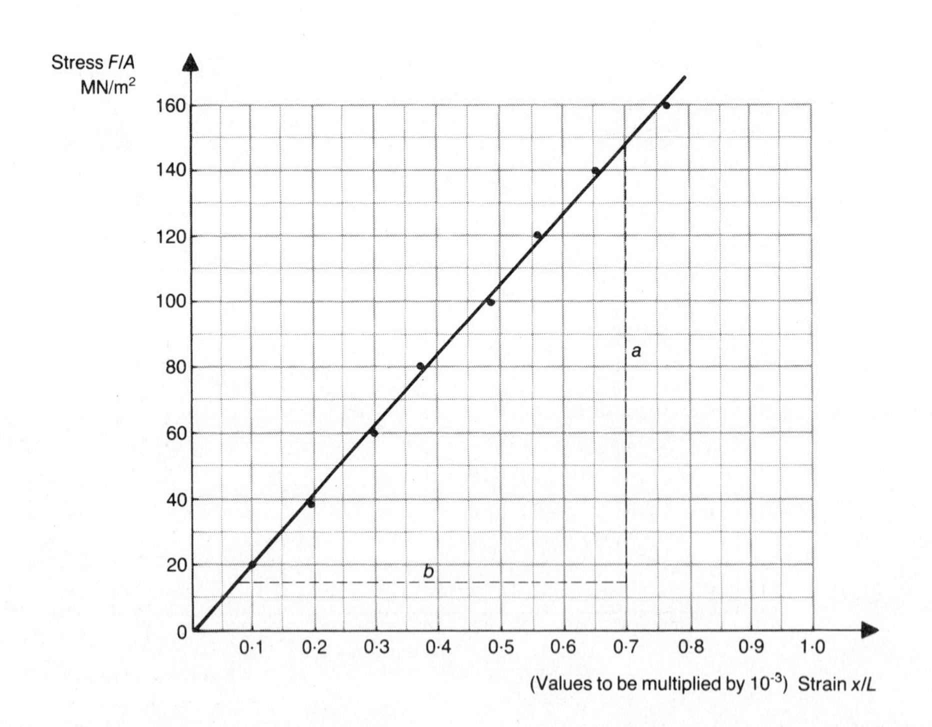

Figure 2.7 Plot of stress versus strain over Hooke's law linear range to determine Young's modulus of elasticity

.4 Practical problems involving Hooke's law and elasticity

xamples

1 A tensile force of 50 N is applied to a brass wire of length 1.2 m and produces an extension of 0.5 mm. Assuming that Hooke's law is obeyed, determine:
(a) the extension for a tensile force of 40 N;
(b) the force required to produce an extension of 0.8 mm;
(c) the extension if the length of the wire is reduced to 0.8 m and the load applied is 25 N.

olution

(a) Since Hooke's law is obeyed, we have

$$x \propto F \quad \text{or} \quad x = kF$$

where x = extension, F = tensile force and k is a constant of proportionality.

When $F = 50$ N, extension $x = 0.5$ mm, so on substituting these values in the formula $x = kF$, we have

$$0.5 = 50\,k$$
$$\text{so} \quad k = 0.5/50 = 0.01$$

Hence $F = 40$ N, $x = 0.01 \times 40 = 0.4$ mm *Ans*

(b) Substituting $x = 0.8$ mm into the formula

$$x = 0.01\,F$$
$$\text{we have} \quad 0.8 = 0.01\,F$$
$$\text{so } F = 0.8/0.01 = 80 \text{ N} \quad \textit{Ans}$$

i.e. a force of 80 N will produce an extension of 0.8 mm

(c) If the 25 N load is applied to the original 1.2 m wire length, the extension produced is

$$x = 0.01 \times 25 = 0.25 \text{ mm}$$

Now as stress (F/A) is directly proportional to strain (x/L)

i.e. stress = $E \times$ strain

and as the same 25 N load is applied to both 1.2 m and 0.8 m lengths of wire, both wires experience the same stress and so the strains produced must also be equal.

$$\text{Therefore } \frac{0.25 \text{ mm}}{1.2 \text{ m}} = \frac{x_1 \text{ mm}}{0.8 \text{ m}}$$

where x_1 = extension produced in 0.8 m wire

$$\text{Thus } x_1 = \frac{0.25}{1.2} \times 0.8 = 0.167 \text{ mm} \quad \textit{Ans}$$

2 A tensile force of 50 kN (50×10^3 N) when applied to a metal rod of length 0.5 m and area 9×10^{-4} m^2 produces an extension of 0.29 mm. Assuming that Hooke's law applies, calculate:
(a) the extension produced by a load of 75 kN;
(b) the value of the critical load corresponding to the elastic limit if it is found that the rod does not return to its original length when an extension of 0.5 mm is exceeded;
(c) Young's modulus for the rod material.

Solution

(a) As extension is directly proportional to applied force, when the load is increased from 50 kN to 75 kN the extension x will be increased by the same ratio, i.e.

$$x \propto 75 \quad \text{as} \quad 0.29 \propto 50$$

$$\text{so } x = 0.29 \times \frac{75}{50} = 0.435 \text{ mm} \quad \textit{Ans}$$

(b) The value of load corresponding to the elastic limit corresponds to an extension of 0.5 mm, i.e.

$$0.5 \propto F$$
$$\text{but } 0.29 \propto 50 \text{ kN}$$

$$\text{so } F = 50 \times \frac{0.5}{0.29} = 86.21 \text{ kN} \quad \textit{Ans}$$

(c) Young's modulus, $E = \dfrac{\text{stress}}{\text{strain}}$

for loads below the elastic limit.

Thus evaluating stress and strain for the case of the 50 kN load, we have

$$\text{stress} = \frac{F}{A} = \frac{50 \times 10^3}{9 \times 10^{-4}}$$
$$= 5.56 \times 10^7 \text{ N/m}^2$$

$$\text{strain} = \frac{x}{L} = \frac{0.29 \times 10^{-3}}{0.5} = 5.8 \times 10^{-4}$$

$$\text{so } E = \frac{5.56 \times 10^{7}}{5.8 \times 10^{-4}} = 9.58 \times 10^{10}\,\text{N/m}^2$$

Ans

3 Cable X consists of a material of tensile strength (maximum stress it can withstand without breaking) of 400 MN/m^2, while the material of cable Y has a tensile strength of 150 MN/m^2. The cross-sectional areas of cables X and Y are 80 mm^2 and 250 mm^2, respectively.
(a) Calculate the maximum load that each cable could support.
(b) For safety reasons the cables cannot be used for loads in excess of one-tenth of the maximum as determined by their tensile strength and cross-sectional area, i.e. a factor of safety of 1/10 is used. Assuming that the gravitational force of a mass M kilograms is 10 M newtons, determine the maximum masses that each cable could safely support.

Solution

(a) Since stress = force/area, force = stress × area. We are given the maximum stress for both cases:

cable X, maximum stress
= tensile strength

so maximum load
= tensile strength × cross-sectional area in m^2
= $(400 \times 10^6) \times (80 \times 10^{-6})$
= 32 000 N = 32 kN *Ans*

(Remember: 1 mm^2 = $10^{-3} \times 10^{-3} = 10^{-6}$ m^2, and 1 MN = 10^6 N)

cable Y, maximum load
= $(150 \times 10^6) \times (250 \times 10^{-6})$
= 37 500 N = 37.5 kN *Ans*

(b) Using a factor of safety of one-tenth, the maximum loads cables X and Y can support are, respectively:

$F_X = 32/10 = 3.2\,\text{kN} = 3200\,\text{N}$
$F_Y = 37.5/10 = 3.75\,\text{kN} = 3750\,\text{N}$

and since we are given the force due to gravitation on a mass is 10 M, the corresponding masses that each cable can safely support are:

$F_X = 10\,M_X$, so $M_X = 3200/10 = 320\,\text{kg}$ *Ans*

$F_Y = 10\,M_Y$, so $M_Y = 3750/10 = 375\,\text{kg}$ *Ans*

2.5 Properties of materials: brittle, malleable and ductile materials

A brittle material is one which when subjected to tensile, compressive, or shear stress can undergo only a relatively small strain (typically a few per cent or less) before breaking. Thus if a brittle material experiences high intensity forces (e.g. a strong hammer blow) it is likely to fracture.

A malleable material on the other hand is one which has the capability of being rolled or hammered into thin sheets or various shapes without fractures occurring. For example, gold is extremely malleable and may be rolled into sheets of thickness as small as 0.001 mm without tearing. Lead, copper, tin, and aluminium also possess good malleability and hence may be subjected to relatively large strain without fracture. They withstand high intensity forces (naturally, within certain limits) without fractures occurring. Their ability to be 'shaped' dictates their use in a wide variety of applications both industrially and domestically. Pots, pans, and foil are examples of the latter use.

Ductile materials have the property that they can undergo large tensile strain before they eventually fracture. They possess the ability to be readily elongated or drawn out under tensile strain. For example, mild steel can be extended in excess of 25% of its unstressed length, while platinum can be drawn out into wire as fine as 0.001 mm diameter before breaking. Examples of materials with good ductile properties are structural steel, aluminium, copper, gold, silver, platinum, and glass (when heated to the molten state)

Brittle materials on the other hand can undergo only a very small tensile strain before breaking. Cast iron, for example, fractures under tensile strain of less than 1%. Other examples of brittle materials are chalk, concrete, high-carbon steels (steels containing more than 1% carbon are brittle, whereas steels containing less than 0.5% carbon possess good ductile properties), and glass (at normal temperatures).

The following experimental results were obtained on load–extension tests on a ductile material (mild steel) and a brittle material (cast iron). To illustrate the difference between their properties in the tensile testing, percentage strain defined as

$$\text{Percentage strain} = \frac{\text{extension, } x}{\text{unstressed length, } L} \times 100$$

has been used in the load–extension graphs shown in Figure 2.8. These graphs were plotted using the following experimentally determined results taken on a tensile-testing bench in an engineering laboratory.

Ductile material (mild steel)

Load F (N)	Extension x (mm)	% Strain $\frac{x}{L} \times 100$
0	0	0
5×10^3	0.013	0.024
10×10^3	0.027	0.047
15×10^3	0.040	0.071
20×10^3	0.062	0.11
25×10^3	0.090	0.16
30×10^3	0.158	0.28
30×10^3	0.452	0.80
30×10^3	1.017	1.80
32.5×10^3	1.582	2.80
35×10^3	2.03	3.60
37.5×10^3	2.49	4.40
40×10^3	3.16	5.60
42.5×10^3	4.52	8.00
45×10^3	5.65	10.00
7.5×10^3	7.91	14.0
49×10^3	10.17	18.0
45×10^3	13.90	24.6
7.5×10^3	15.82	28.0

Brittle material (cast iron)

Load F (N)	Extension x (mm)	% Strain $\frac{x}{L} \times 100$
0	0	0
5×10^3	0.023	0.041
10×10^3	0.047	0.083
12.5×10^3	0.068	0.12
15×10^3	0.085	0.15
17.5×10^3	0.113	0.20
18.5×10^3	0.130	0.23

Physical data for both specimens: unstressed length L = 56.5 mm, radius r = 5.65 mm, cross-sectional area $A = \pi r^2$ = 100 mm^2 = 10^{-4} m^2.
To convert load F to stress divide by cross-sectional area, in effect × F by 10^4.
To convert extension x to strain divide by L.
By international convention length of specimen, $L = \sqrt{A}$ where A = cross-sectional area.

Let us examine first the graph of Figure 2.8(a) for mild steel which has a number of distinctive regions fairly typical of the load–strain characteristics of a ductile material:

Range OA: over the range of strain 0 to 0.08% the extension is proportional to load. At point A where the load is 15×10^3 N the graph deviates from a straight line, and thus A defines the limit of proportionality and OA is the region in which Hooke's Law is obeyed, i.e. for 5×10^3 N load increases up to 15×10^3 N, the extension increases by equal amounts.
Range AB: between A and B the load–extension graph is curved and just above point A the elastic limit is exceeded. Up to this point the test specimen will return to its original unstressed length if the load is removed. After the elastic limit point a permanent extension will be produced in the material when the load is removed.
Range BC: at point B corresponding to a load of 30×10^3 N a sudden increase in extension occurs with little or no increase in load to point C. Point B is known as the yield point, and plastic-type deformation occurs between B and C.
Range CD: over this range uniform plastic deformation occurs, i.e. elongation of the

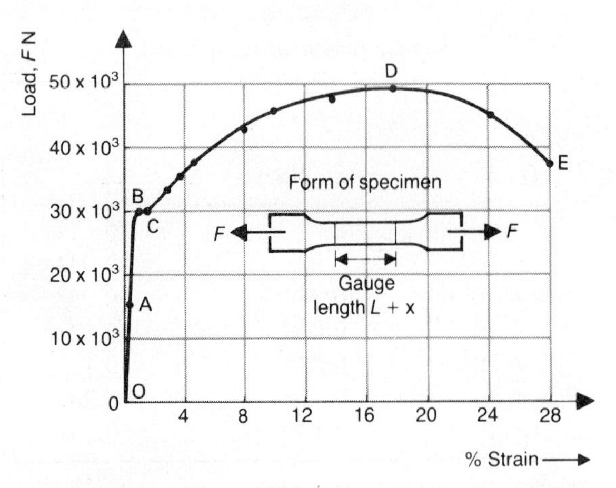

(a) Load versus percentage strain for mild steel (ductile material)

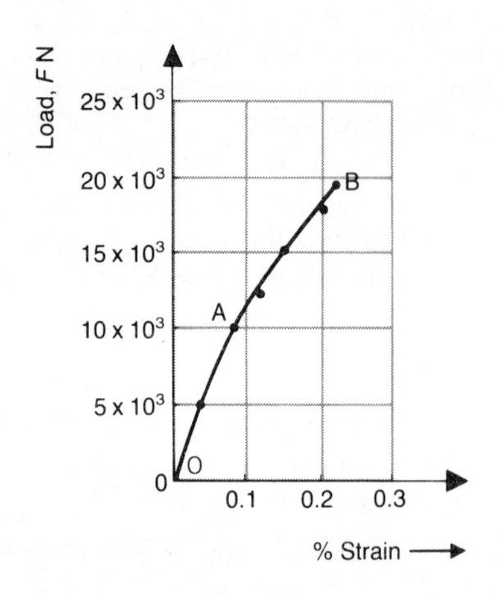

(b) Load versus percentage strain for cast iron (brittle material)

Figure 2.8
(a) Load versus percentage strain for mild steel (ductile material).
(b) Load versus percentage strain for cast iron (brittle material)

specimen occurs combined with uniform reduction of its cross-sectional area. D, corresponding to a load of 49×10^3 N, is the maximum load beyond which the specimen extends and finally fractures. The stress at this point,

$$\frac{F_D}{A} = \frac{49 \times 10^3}{10^{-4}} = 490 \times 10^6 \text{ N/m}^2$$

is the maximum stress the material can withstand and is known as the tensile strength.

Range DE: after D is passed, localised plastic deformation takes place, in which a 'neck' is formed in the specimen. This results in the fact that an increase in stress can be obtained with a reduction in load and thus the breaking load at E is actually smaller than D.

The graph of Figure 2.8(b) for cast iron typifies the load–strain characteristics of a brittle material. Over the range OA the material obeys Hooke's law with the extension and strain directly proportional to the applied load. Over the range AB there is a small amount of plastic deformation but no yield point. Abrupt fracture occurs at B, corresponding to a load of just under 20 kN and a strain of only 0.23%.

Test 2

This test may be used as a basic self-assessment test to check whether you have absorbed the main facts of the second chapter on **Elasticity and Hooke's law** and its objectives.

Qu. 1 Enter a tick (√) in the answer block if you consider the statement correct; enter a cross (×) if you consider the statement incorrect, even though part of it could be right.

(a) Materials which are subjected to forces which exceed their elastic limit will not return to their original unstressed dimensions when the forces are removed.

(b) Most metals are elastic and obey Hooke's law for applied forces which cause strains of approximately 0.1% or less.

(c) Stress is force per unit area.

(d) A tensile force of 50 N acting on a rod of given material of cross-sectional area 10 mm² produces a greater tensile stress than a force of 10 N acting on a rod of the same material but of smaller area of 0.1 mm².

(e) Hooke's law states that strain is directly proportional to stress.

Answer block:

<table>
<tr><td rowspan="2">Question no.</td><td colspan="5">1</td><td rowspan="2">2</td></tr>
<tr><td>(a)</td><td>(b)</td><td>(c)</td><td>(d)</td><td>(e)</td></tr>
<tr><td rowspan="5">Enter answer →</td><td rowspan="5"></td><td rowspan="5"></td><td rowspan="5"></td><td rowspan="5"></td><td rowspan="5"></td><td>(a)</td></tr>
<tr><td>(b)</td></tr>
<tr><td>(c)</td></tr>
<tr><td>(d)</td></tr>
<tr><td>(e)</td></tr>
</table>

<table>
<tr><td rowspan="2">Question no.</td><td colspan="4">3</td><td colspan="3">4</td></tr>
<tr><td>(a)</td><td>(b)</td><td>(c)</td><td>(d)</td><td>(a)</td><td>(b)</td><td>(c)</td></tr>
<tr><td>Enter answer →</td><td></td><td></td><td></td><td></td><td></td><td></td><td></td></tr>
</table>

Qu. 2 Write in the answer block the most appropriate word(s) missing in the following statements so they are complete and make sound scientific sense.

(a) The force of a roof on a load-bearing wall produces ... stress in the wall.

(b) When a force acts parallel to the cross-sectional area of a material it produces ... stress.

(c) Brittle materials can undergo only relatively small strain, whereas ... materials such as mild steel can undergo relatively large strain before eventual fracture.

(d) If the tensile loading of a material is doubled so is the ... provided Hooke's law is obeyed.

(e) The ratio of stress to strain over the linear range of a material subjected to tensile force is a ... and is known as ....

Qu. 3 A load of $F = 100$ N produces an extension of 1 mm in a wire of unstressed length 2 m. Assuming that Hooke's law is obeyed, calculate:

(a) the extension if the load is reduced to 60 N;

(b) the load which would cause an extension of 1.5 mm;

(c) the extension if the wire is reduced in length to 1 m and then subjected to an 80 N load;

(d) the extension produced in a wire of the same material, 1 m in length but double the cross-sectional area, when loaded by a force of 100 N.

Qu. 4 The following results were obtained on loading a light spring.

load F (N)	0	20	40	60	80
extension x (mm)	0	5.5	10.9	16.0	21.5
load F (N)	100	120	140	160	
extension x (mm)	26	31	39	52	

Plot a graph of extension versus load and use the graph to determine:

(a) the extension that would be produced by a load of 65 N;

(b) the load that produces an extension of 22 mm;

(c) the maximum load corresponding to the limit of the linear Hooke's law region.

Problems 2

1 (a) Define what is meant by an elastic material and state Hooke's law.
(b) Describe an experiment to investigate the relationship between (i) a metal wire; (ii) a strip of rubber of uniform cross-section, when they are subjected to a tensile force.
Sketch graphs of extension versus force for both cases, illustrating the results you might expect.

2 A metal bar of 1 m length and cross-sectional area 5 mm^2 is extended 2 mm when subjected to a tensile force of 220 N. Calculate:
(a) the tensile stress in the bar;
(b) the extension if the tensile force is increased to 350 N, assuming Hooke's law applies;
(c) the applied force necessary to extend the bar by 1.5 mm.

3 (a) Define stress and strain.
(b) A tensile force of 75 kN (75×10^3 N) produces an extension of 0.4 mm in a metal bar of unstressed length of 0.4 m and cross-sectional area of 50 mm^2. Calculate, assuming that Hooke's law applies:
(i) the extension produced by a force of 100 kN;
(ii) the tensile stress in the bar when subjected to a force which produces an extension of 0.5 mm.

4 A steel cable of cross-sectional area 200 mm^2 carries a load of 10 kN. Calculate the tensile stress in the cable.
If the maximum allowable working stress is 80 MN/m^2, determine the minimum cross-sectional area of the cable that could safely be used to support the 10 kN load.

5 The following experimental results were obtained from a tensile loading test on a length of steel wire:

Tensile force F (N)	0	10	20	30	40
Extension x (mm)	0	0.24	0.6	0.84	1.10

Tensile force F (N)	50	60	70	80
Extension x (mm)	1.41	1.68	1.93	2.25

Plot a graph of extension versus tensile force and comment on whether the results show that Hooke's law applies. Determine from the graph:
(a) the tensile force which produces an extension in the wire of 0.75 mm;
(b) the extension produced by a tensile force of 66.5 N.

6 Describe how you would investigate experimentally the relationship between the extension of a light spring when subjected to varying loads.
Explain the meaning of the term 'elastic limit' and, assuming that the spring is not loaded beyond this limit, sketch the graph of extension versus load that you would expect from the results of your experiment.
Describe also how you would determine the mass of an unknown weight which does not load the spring beyond the elastic limit.

7 Given that the tensile strength of a steel cable of cross sectional area 300 mm^2 is 450 MN/m^2 estimate the maximum load that the cable could support.

3 Forces, moments and static equilibrium

General learning objectives: to establish the effect of a force rotating about a point and to solve problems relating to static equilibrium and finding the resultant force.

3.1 Introduction: statics, equilibrium, coplanar forces

Statics is a branch of mechanics relating to the mathematical and physical study of the behaviour of materials and bodies under the action of forces where no motion is produced – hence the term statics. In Chapter 2, on elasticity and Hooke's law, we considered the properties of material in which the applied forces produced stress. No motion was involved since in equilibrium the applied force was always counterbalanced by equal and opposite stress forces set up in the material.

In this chapter we continue our study of statics by considering forces acting on bodies and the conditions necessary for equilibrium, equilibrium being a state of balance between opposing forces such that the body on which they are acting remains at rest. The resultant force and turning moment are zero. We restrict our study to the coplanar case, coplanar meaning 'in the same plane'. Thus coplanar forces are forces which act in the same plane. Figures 3.1(a) and (b) show examples of coplanar forces; the forces in (c) are not coplanar, they act in three dimensions.

3.2 Scalar and vector quantities

A **scalar** quantity is one which is sufficiently defined when the magnitude only is given, in appropriate units. For example, the following quantities are scalar:

mass e.g. 10 kg expresses the mass of a body;

temperature e.g. 0°C, 68.2°C are sufficient to define the temperature of a body;

time e.g. 12 s, 6 hours, 3 days, 10 light-years, all express the 'magnitude' of time.

A **vector** quantity, however, is one which requires both a magnitude and a direction to be stated to define it completely. Examples of vector quantities include:

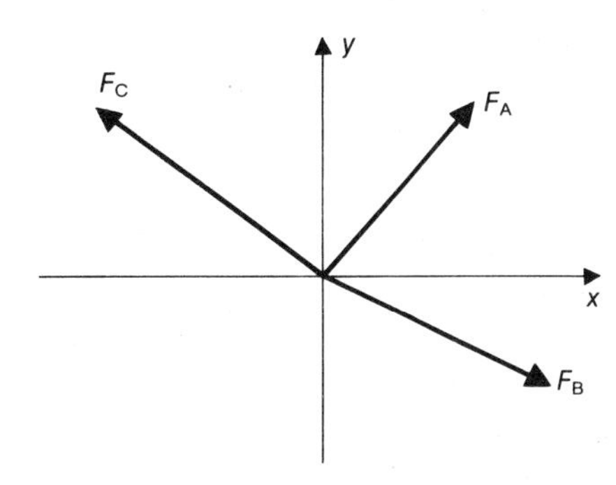

(a) Forces F_A, F_B and F_C are coplanar, they act in the same plane

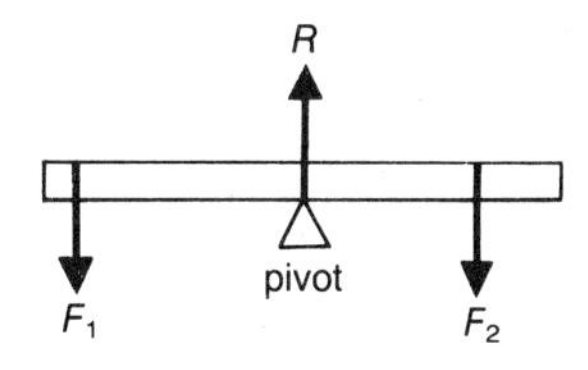

(b) Forces F_1, R and F_2 are coplanar, they all act in the same vertical plane

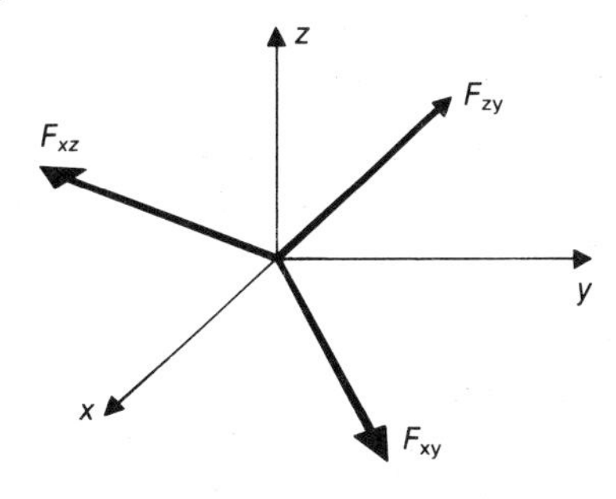

(c) Forces F_{xy}, F_{yz} and F_{xz} are not coplanar

Figure 3.1

force – both its magnitude and the direction in which it acts must be defined;
velocity – the magnitude (speed) and the direction must be given to fully define velocity;
acceleration (rate of change of velocity) – both the magnitude and the direction should be specified;
electric and magnetic fields – again both the strength and the direction must be given to fully define field.

Vector quantities may be represented on a diagram by means of a straight line of a given length and drawn in a given direction. The length of the line represents the magnitude of the quantity and the line direction, usually marked with an arrowhead, the direction in which the quantity acts. For example, a velocity v_1 of 1.6 m/s in the direction of a fixed line AN may be represented on a diagram by drawing a line of length 16 mm (scale: 10 mm represents 1 m/s) in the direction AN as shown in Figure 3.2(a); (b) represents a velocity v_2 of magnitude 2.4 m/s and in a direction at 30° to AN; and (c) represents a velocity v_3 of magnitude 0.85 m/s in a direction at 90° to AN. A convenient means of specifying vectors is to write down their magnitude followed by their angle to a given fixed direction. Thus the velocities in Figure 3.2 may be written as:

$v_1 = 1.6\,\text{m/s}\ \angle 0°$
↑ direction with respect to AN
magnitude of velocity

$v_2 = 2.4\,\text{m/s}\ \angle 30°$, magnitude = 2.4 m/s, direction at 30° to AN

$v_3 = 0.85\,\text{m/s}\ \angle 90°$, magnitude = 0.85 m/s, direction 90° to AN.

A force is fully defined by giving:

1 its magnitude in newtons (N), the newton is the SI unit of force;
2 its line of action, i.e. the direction in which the force 'pulls' or 'pushes';
3 its point of application.

For example, in Figure 3.3 four forces F_1, F_2, F_3, and F_4, acting at point O, have been drawn. Point O is the point of application. A scale of 10 mm = 20 N (i.e. 1 mm represents 2 N) is used and line OX is chosen as the reference direction,

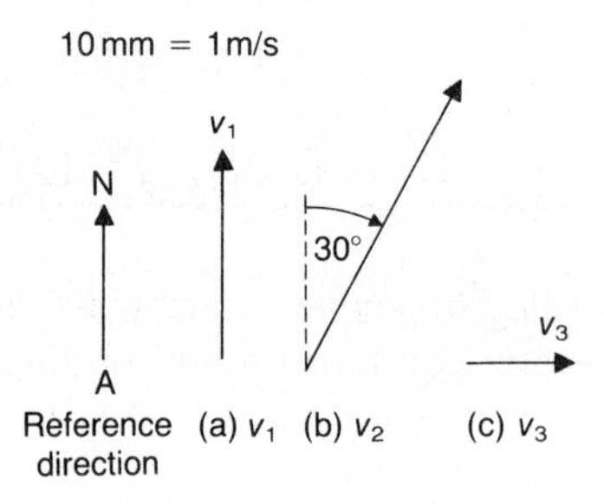

Figure 3.2 Velocity vectors

e.g. a force of 20 mm length in direction OX would represent the force 40 N $\angle 0°$. Using a ruler and protractor determine the magnitude and direction of the four forces. The answers are:

$F_1 = 38\,\text{N}\ \angle 90°$ (length OA = 19 mm, so magnitude of F_1 is $20 \times 19 \div 10 = 38\,\text{N}$, F_1 makes $\angle AOX = \angle 90°$, with reference direction OX);
$F_2 = 37\,\text{N}\ \angle 27°$; $F_3 = 40\,\text{N}\ \angle 315°$;
$F_4 = 32\,\text{N}\ \angle 180°$.

Note that all the angles associated with specifying the direction of the forces are measured from OX in the counter-clockwise direction. By convention this direction is taken as positive. Angles measured in the clockwise direction are taken as negative angles. Thus F_3 may be quoted, as above, as 40 N $\angle 315°$, or alternatively as 40 N $\angle -45°$.

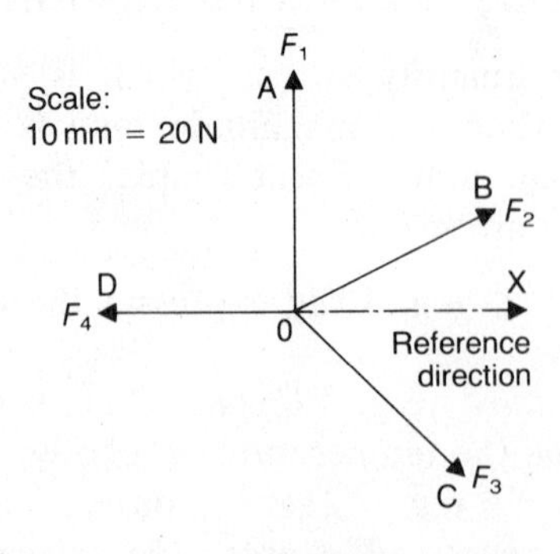

Figure 3.3 Coplanar forces acting at point 0

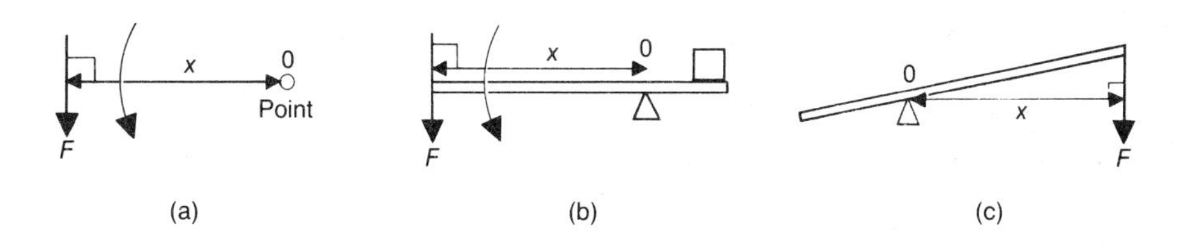

Figure 3.4 Moment of a force = force × perpendicular distance = F × x newton-metres (Nm)

3.3 The principle of moments

In this section we consider the 'turning' effect produced by forces and the conditions for equilibrium as embodied in the principle of moments. First let us define the term 'moment':

> *The moment of a force about a point is the product of the magnitude of the force and the perpendicular distance of the line of action of the force to the point.*

Referring to Figure 3.4,

moment of force F about point O $= F \times x$

where x = perpendicular distance of line of action of F from O. The units of moment are newton-metres (Nm). In Figure 3.4(a) and (b) the moment is in a counter-clockwise direction, in (c) the moment is clockwise.

A body is in equilibrium, provided the following two conditions are satisfied:

1. The resultant of the forces acting on the body is zero.
2. The sum of the clockwise moments about any point equals the sum of the counter-clockwise moments about the same point.

When condition 1 is satisfied there is no translational motion, i.e. no movement, for example, in a straight line. When condition 2 is satisfied there will be no rotational or turning movement. Condition 2 is the statement of the *principle of moments*. This principle is applied to solve equilibrium problems when we are dealing with forces which do not all act at a common point.

The application of the equilibrium conditions and in particular the principle of moments is illustrated in the following examples.

Examples

1 Forces $F_1 = 25\,N$, $F_2 = 50\,N$ act vertically upwards on a light rod (whose mass can be neglected), force $F_3 = 75\,N$ acts vertically downwards as shown in Figure 3.5. Show that the rod is in equilibrium.

Solution

The upward forces F_1 and F_2 are balanced by F_3 acting downwards, as

$$F_1 + F_2 = 25 + 50 = 75\,N \text{ and } F_3 = 75\,N$$

We also check that there is no rotational motion by applying the principle of moments. We can select any point about which to take moments, so let us first try the centre point M on the rod. If we consider the rod as pivoted at M, then forces F_1 and F_3 will tend to rotate the rod in a clockwise direction, while F_2 will cause a rotation in the counter-clockwise direction. For the rod to be in equilibrium the sum of the clockwise moments must be exactly counterbalanced by the counter-clockwise moment, i.e.

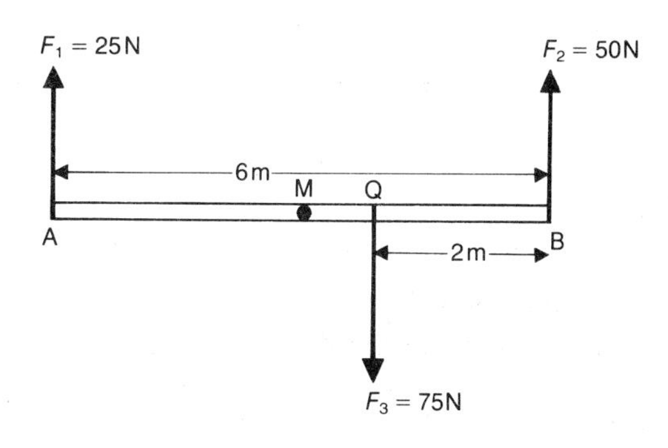

Figure 3.5 Diagram for example 1

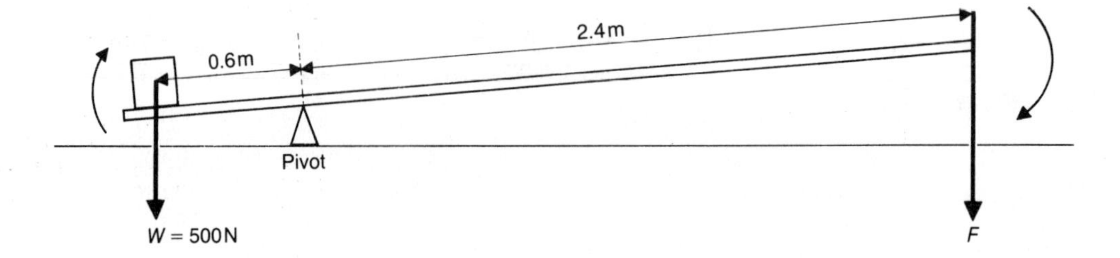

Figure 3.6 Diagram for example 2

Clockwise moments about M
$= (F_1 \times \text{AM}) + (F_3 \times \text{MQ})$
$= (25 \times 3) + (75 \times 1) = 150\,\text{Nm}$
as AM = 3 m, MQ = 1 m
Counter-clockwise moment about M
$= F_2 \times \text{MB}$
$= 50 \times 3 = 150\,\text{Nm}$

Thus the total clockwise moment equals the counter-clockwise moment and the rod is in equilibrium.

Suppose we took moments about end A, the moment due to F_1 would then be zero and we would obtain

$F_3 \times \text{AQ} = 75 \times 4 = 300\,\text{Nm}$ clockwise
$F_2 \times \text{AB} = 50 \times 6$
$= 300\,\text{Nm}$ counter-clockwise

The two moments are again equal in magnitude but in opposing direction and so balance. We need only apply the principle of moments about one point. If it holds for one point it will also be satisfied for any other. If it fails for one point, it will fail for all others and equilibrium cannot occur.

2 Figure 3.6 shows a diagram of a simple bar lever. Estimate the minimum value of the force F required to raise the load $W = 500\,\text{N}$.

Solution

If the force F is to just raise the load W, its moment about the pivot must equal the counter-clockwise moment of the load force $W = 500\,\text{N}$ about the same point.

Thus $W \times 0.6 = F \times 2.4$

as the 'perpendicular' distances of W and F from the fulcrum are 0.6 m and 2.4 m respectively. A small approximation is made since we assume the bar is horizontal, so W and F act at right-angles to the bar, even though the diagram shows the bar slightly tilted. Substituting $W = 500\,\text{N}$, we have

$$500 \times 0.6 = 2.4F$$

$$\text{so } F = \frac{500 \times 0.6}{2.4} = 125\,\text{N} \quad \textit{Ans}$$

3 A uniform steel beam of length 8 m is supported at each end by vertical pillars. The beam carries loads of 10 kN and 25 kN acting respectively at 2 m and 6 m from the left-hand support. The weight of the beam is 2 kN and may be taken to act vertically downwards at its centre. The complete force diagram is drawn in Figure 3.7. Calculate the reaction forces R_1 and R_2 at the supporting pillars.

Solution

Taking moments about end A and applying the principle of moments we have, with forces in kN units,

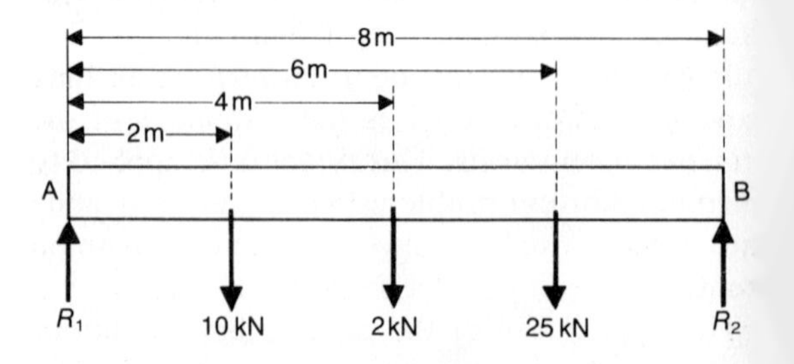

Figure 3.7 Diagram for example 3

$$(10 \times 2) + (2 \times 4) + (25 \times 6) = R_2 \times 8$$

i.e. $20 + 8 + 150 = 8R_2$

$$\text{so } R_2 = \frac{178}{8} = 22.25\,\text{kN} \quad \textit{Ans}$$

Likewise, taking moments about B,

$$(25 \times 2) + (2 \times 4) + (10 \times 6) = R_1 \times 8$$

so $8R_1 = 50 + 8 + 60 = 118$

$$R_1 = \frac{118}{8} = 14.75\,\text{kN} \quad \textit{Ans}$$

Checking also to see that total upward reaction forces equal the total downward forces, we have:

$R_1 + R_2 = 14.75 + 22.25 = 37\,\text{kN}$

Total downward forces $= 10 + 2 + 25 = 37\,\text{kN}$

3.4 Centre of gravity

The **centre of gravity** of a body is that point through which the resultant of the earth's gravitational force of attraction upon the body can be considered as acting.

The total mass of a body is of course distributed throughout its volume. However, as far as the gravitational force on the body is concerned, we can take the total mass to be concentrated at the centre of gravity point. This greatly simplifies force problems when we consider taking into account the weight of bodies. The **weight**, which acts through the centre of gravity, is the force of attraction of the earth on the body and is given by

$W = mg$ newtons

where m = mass of body in kilograms (kg)

g = acceleration due to gravity $= 9.81\,\text{m/s}^2$

(For many practical calculations $g = 10\,\text{m/s}^2$ is often used as an approximation.)

In Figure 3.8 the individual parts of the body all experience downward forces due to gravitational attraction between the parts and the earth. The resultant force of attraction, $W = mg$, acts through the centre of gravity G of the body, so for example, the moment due to W about point P is $mg \times x$, where x is the perpendicular distance from P to the line of action of W acting through G.

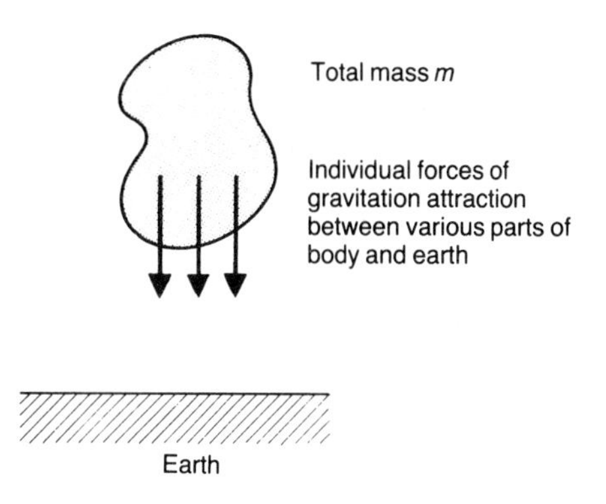

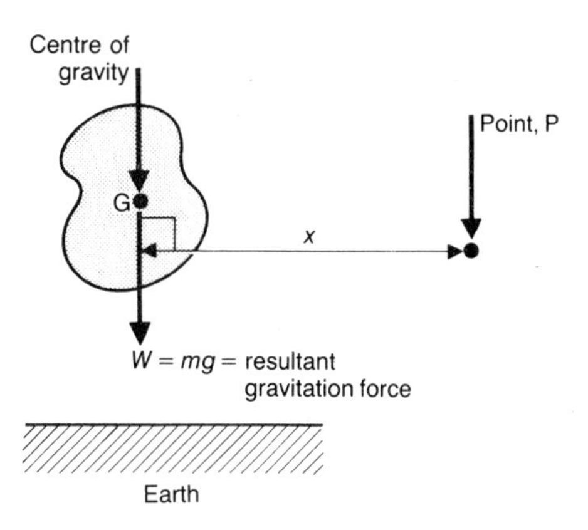

Figure 3.8

The centres of gravity for a uniform rod, a rectangular lamina (a lamina is a thin plate or sheet of material) and a circular lamina are shown in Figure 3.9.

When any body is suspended at a point such that it can move freely it will always come to rest with its centre of gravity lying vertically below the point of suspension. It is only in this way that the principle of moments can be satisfied.

This fact enables us to obtain experimentally the centre of gravity of sheets of material, however irregular their boundary, provided they are of constant thickness.

Consider, for example, the determination of the centre of gravity of a piece of cardboard such as that shown in Figure 3.10(a). Pierce the cardboard with a fine needle at any point fairly close to its boundary and mount the sheet vertically, as shown in (b). Make sure the sheet is free to rotate in the vertical plane. In its equilibrium position its centre of gravity G will lie vertically below the point of suspension X, i.e. on line XGQ. This line can be marked on the

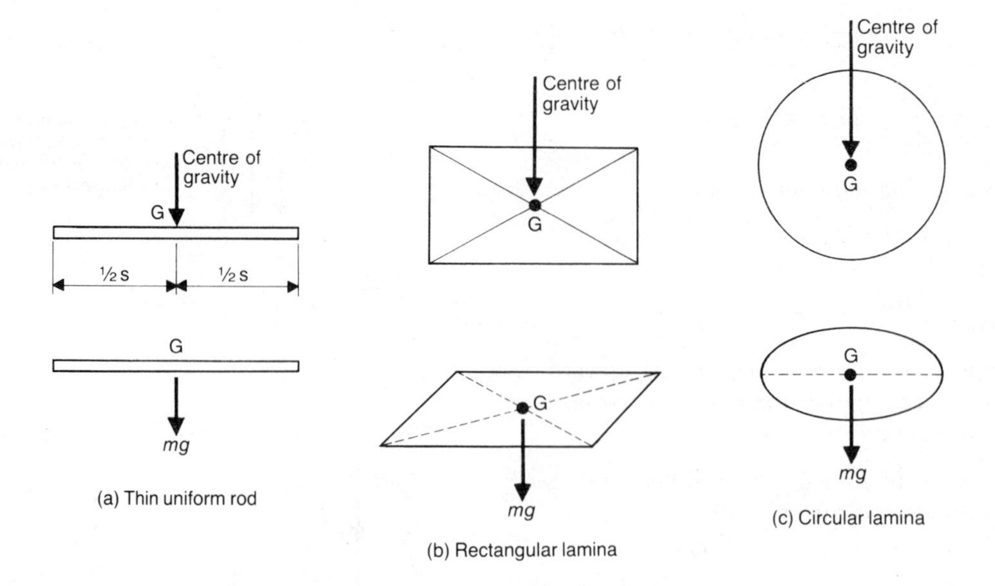

Figure 3.9 Centre of gravity for a thin rod and rectangular laminas

cardboard as follows. From the needle at X suspend a fine length of thread with a small weight attached at its lower end and let it hang vertically very close to the cardboard. Mark two points on the line where the thread 'touches' the cardboard. Draw in the straight line XGQ.

Select a second point Y, again close to the boundary and a good distance from X, as in (c). Repeat the procedure to obtain the straight line YGR. The centre of gravity must lie on this line as well as line XGQ. Thus the centre of gravity G lies at the intersection of the two lines XGQ and YGR.

3.5 Calculation of resultant force: vector diagram techniques

The resultant of two or more forces acting at a point is determined easily when the forces act in the same or opposing directions. Add forces acting in the same direction and subtract forces acting in the opposing direction. For example, in Figure 3.11 the resultant force F acting at point O is

(a) $F = F_1 + F_2 = 30 + 20 = 50\,\text{N}$
(b) $F = F_1 - F_2 = 40 - 25 = 15\,\text{N}$

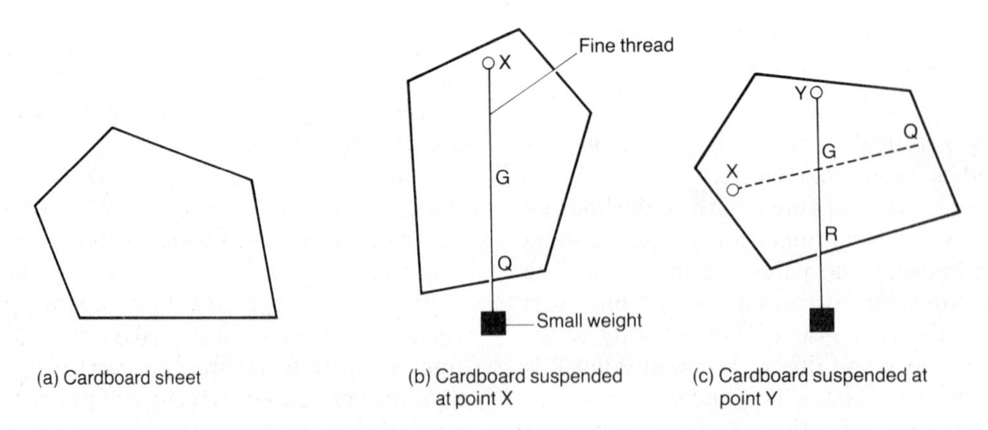

Figure 3.10 Experimental determination of centre of gravity of a uniform sheet of material

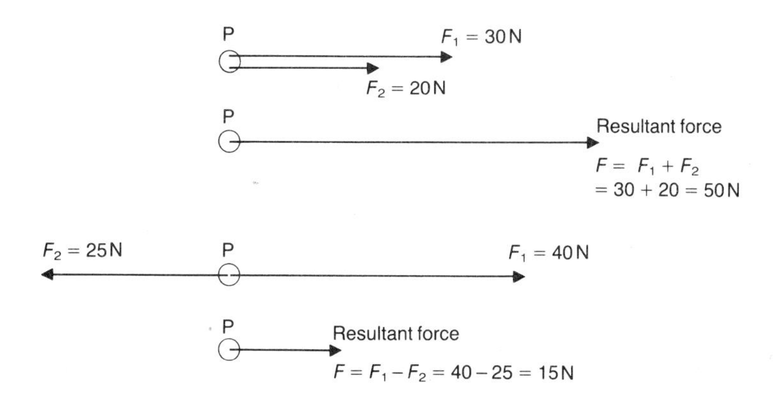

Figure 3.11 Resultant of forces when their lines of action are in the same straight line. (Scale 1 mm: 1 N)

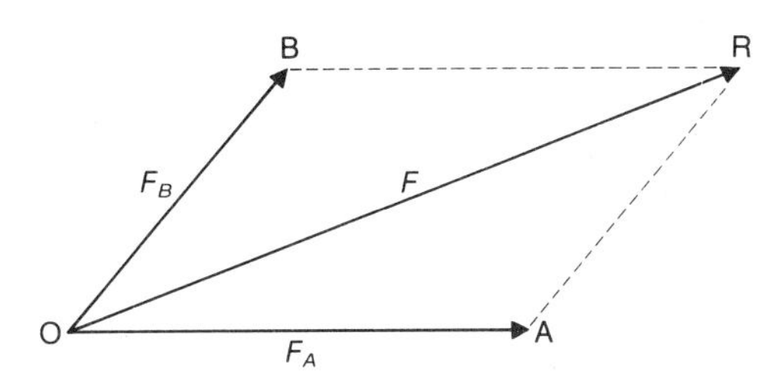

Parallelogram of forces to find the resultant of $F_A = 40$ N and F_B 30 N $\angle 49°$ acting at point O. Scale 1 mm : 1 N. Resultant F = OR $\angle$ ROA, OR = 63.5 mm → 63.5 N, $\angle$ ROA = 20.5°, so $F = 63.5$ N $\angle 20.5°$

Figure 3.12

If two forces acting at a point are not in the same straight line, their resultant may be obtained by drawing, utilizing the parallelogram of forces rule:

If two forces acting at a point are represented in magnitude and direction by two adjacent sides of a parallelogram, the diagonal of the parallelogram which passes through the point of action represents their resultant in magnitude and direction.

The application of this rule is demonstrated in the following examples.

Examples

1 The coplanar forces $F_A = 40$ N $\angle 0°$ and $F_B = 30$ N $\angle 49°$ act at a point O. Determine the resultant force.

Solution

(i) Select scale: 1 mm to represent 1 N, 1 mm: 1 N.

(ii) Draw line OA = 40 mm (40 N) to represent F_A; draw line OB = 30 mm (30 N) at 49° to represent F_B in both magnitude and direction, as shown in Figure 3.12.

(iii) Complete the parallelogram OARB by drawing a line through A parallel to OB, and a second line through B parallel to OA.

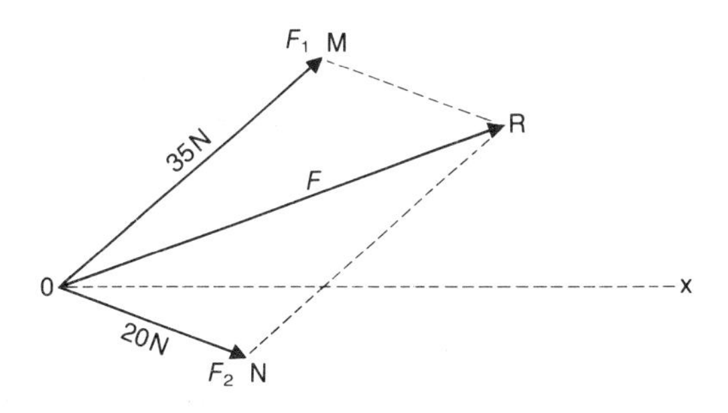

$F_1 = 35$ N $\angle 40°$ (magnitude 35 N at angle of 40° w.r.t. OX)
$F_2 = 20$ N $\angle -20°$ (magnitude 20 N at angle of 20° clock-wise w.r.t. OX)
Resultant of F_1 and F_2 is OR and using scale 1 mm : 1 N we find OR = 48 mm, $\angle$ORX = 19° so $F = 48$ N $\angle 19°$

Figure 3.13

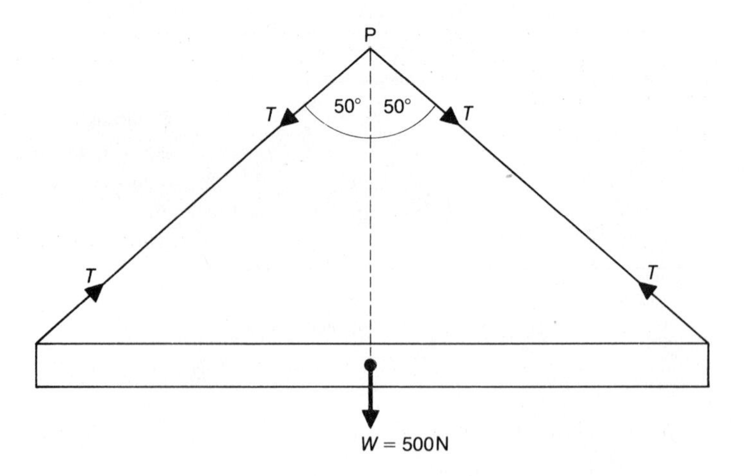

Figure 3.14(a)

(iv) Diagonal OR represents the resultant force. Length OR = 63.5 mm so magnitude of resultant force F = 63.5 N (1 mm = 1 N). ∠ROA = 20.5° which is the angle the resultant force makes with F_A.

Resultant force, F = 63.5 N ∠20.5° *Ans*

2 Find the resultant of the coplanar forces F_1 = 35 N ∠40° and F_2 = 20 N ∠−20°, which act at a common point O in Figure 3.13.

Solution

(i) Draw in reference line OX (0° direction) and select a suitable scale, e.g. 1 mm = 1 N.

(ii) Draw in force F_1 by line OM = 35 mm at an angle of 40° to axis OX.

(iii) Draw in force F_2 by line ON = 20 mm at an angle of −20° with respect to axis OX (i.e. in the clockwise direction or below OX).

(iv) Complete the parallelogram OMRN.

(v) Diagonal OR represents the resultant force,

length OR
= 48 mm, so magnitude = 48 N
∠ROX
= 19°, direction of resultant force, so:
resultant force F = 48 N ∠19° *Ans*

3 A heavy beam of weight 500 N is supported by nylon rope as shown in Figure 3.14(a). The weight of the beam produces tension forces in the rope, indicated by T in the diagram. At point P the resultant force of the two tension forces equal the 500 N weight of the beam acting vertically downwards. Draw a vector diagram at this point to determine T.

Solution

First select a suitable scale, e.g. 1 mm:10 N, and draw in the gravitational force W = 500 N acting vertically downwards, line PR in Figure 3.14(b). At point P draw lines at 50° to vertical to represent the lines of action of the tension forces. Complete the parallelogram by drawing lines through R, i.e. draw in lines RK and RL.

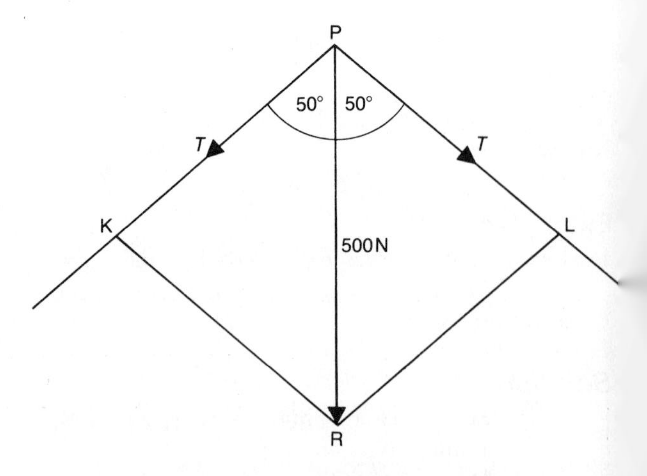

Figure 3.14(b) Vector diagram for example 3
Scale 1 mm: 10 N

Measure PK and PL and convert to newtons. These lines represent the magnitude and directions of the tension force

PK = PL = 39 mm
so T = 390 N *Ans*

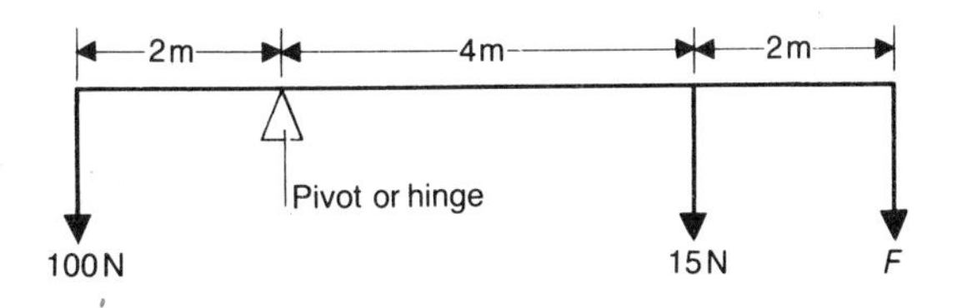

Figure 3.15 Diagram for Qu 2

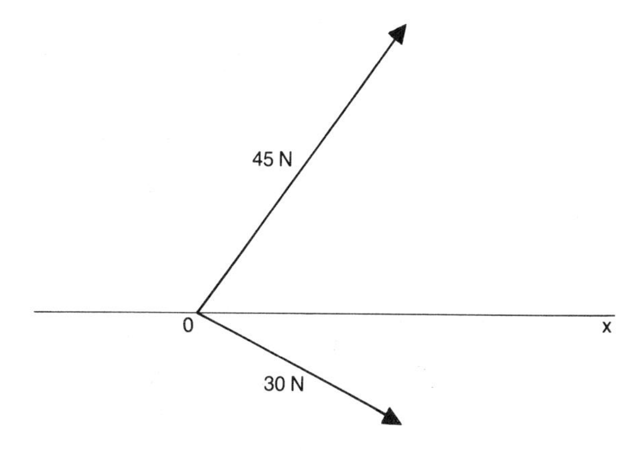

Figure 3.16 Force diagram for Qu 3
Scale: 1 mm: 1 N

Test 3

This test may be used as a basic self-assessment test to check whether you have absorbed the main facts of the third chapter on **Forces, moments and static equilibrium**. All answers to be entered in the answer block.

Qu. 1 Enter a tick (√) in the answer block if you consider the statement correct; enter a cross (×) if you consider the statement in any way incorrect.

(a) 10°C, 52.6 kg, 30 ns are all examples of scalar quantities.
(b) Volume, mass and density are all scalar quantities.
(c) A vector quantity is defined by giving its direction and magnitude.
(d) Velocity, force, length and heat are *all* vector quantities.
(e) The moment of force about a point equals the force magnitude times the perpendicular distance of the line of action of the force from the point.
(f) The centre of gravity of a body is the point at which the total mass can be considered as concentrated and through which the earth's gravitational force of attraction acts.

Qu. 2 Calculate the force F in Figure 3.15 to maintain equilibrium.

Qu. 3 The forces of magnitudes 45 N and 30 N are drawn to scale (1 mm: 1 N) and in the direction they act in the diagram of Figure 3.16. Complete the parallelogram of forces and determine the magnitude and direction (with respect to OX) of the resultant force.

Qu. 4 A uniform beam of length 10 m is pivoted at 2 m from one end as shown in Figure 3.17. Equilibrium is established by attaching a weight of 960 N at this end, also shown. Determine:

Answer block:

Question no.	1						2	3	4
	(a)	(b)	(c)	(d)	(e)	(f)			
Enter answer									

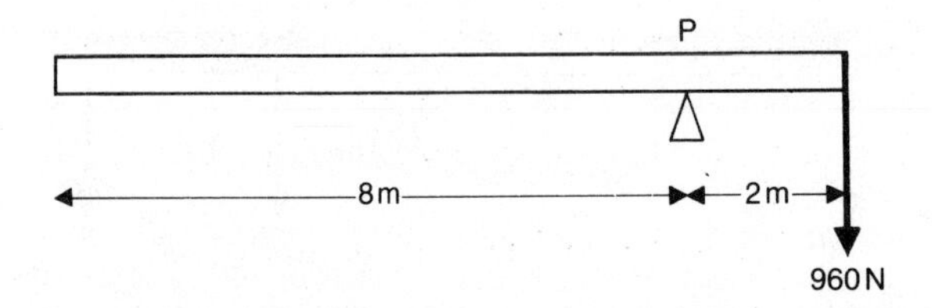

Figure 3.17 Diagram for Qu 4

(a) the distance of the centre of gravity of the beam from the pivot point P;
(b) the weight of the beam in newtons;
(c) the mass of the beam, assuming $g = 10\,\text{m/s}^2$.

Problems 3

1 (a) Distinguish between scalar and vector quantities and give three examples of each.
(b) Define the moment of force about a point and state the principle of moments.

2 Define the centre of gravity of a body and show with the aid of a diagram the position of the centre of gravity for:
(a) a uniform beam;
(b) a rectangular lamina
(c) a solid sphere;
(d) a circular disc.

3 (a) Calculate the magnitude of the force F in the simple lever system of Figure 3.18(a) which will just raise the 1000 N load.

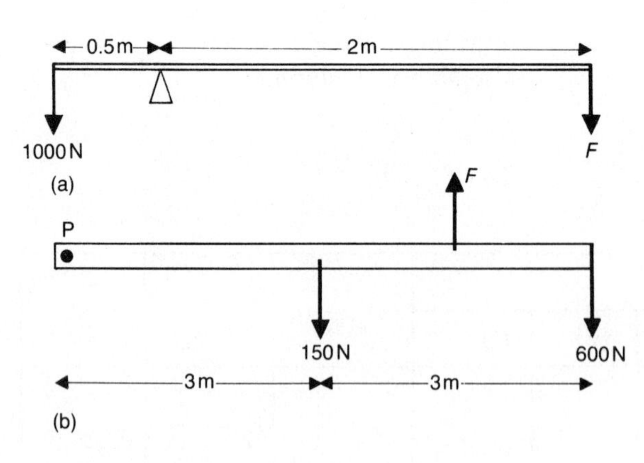

Figure 3.18 For problem 3

(b) The rod shown in Figure 3.18(b) is pivoted at end P and has a weight of 150 N, shown marked in. A load of 600 N is suspended at the far end, 6 m from P. Calculate the magnitude and position of the force F which establishes equilibrium.

4 A uniform beam of mass 50 kg and length 10 m is supported at each end by vertical pillars. The beam carries loads of 300 kg and 600 kg acting at 4 m and 8 m, respectively, from the left end support. Calculate the reaction force at each pillar, assuming g is taken as $10\,\text{m/s}^2$.

5 The diagram of Figure 3.19 shows the disposition of forces acting on a uniform beam suspended by two cables. Determine the tension forces T_1 and T_2 in the cables.

6 State the parallelogram of forces rule for finding the resultant of two forces acting at a point. Apply the rule to find the resultant of the two forces shown in the diagram of Figure 3.20.

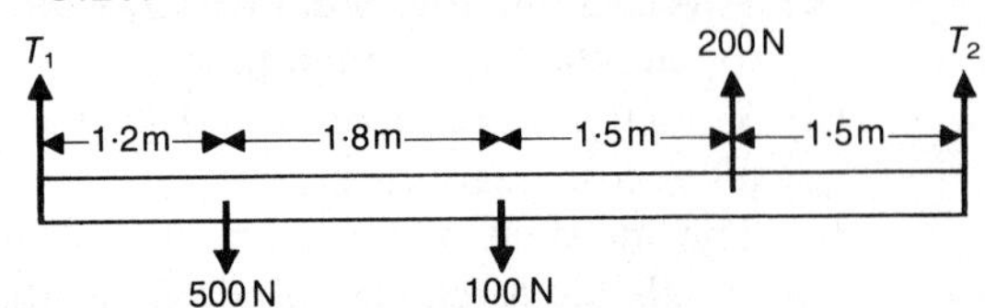

Figure 3.19 For problem 5

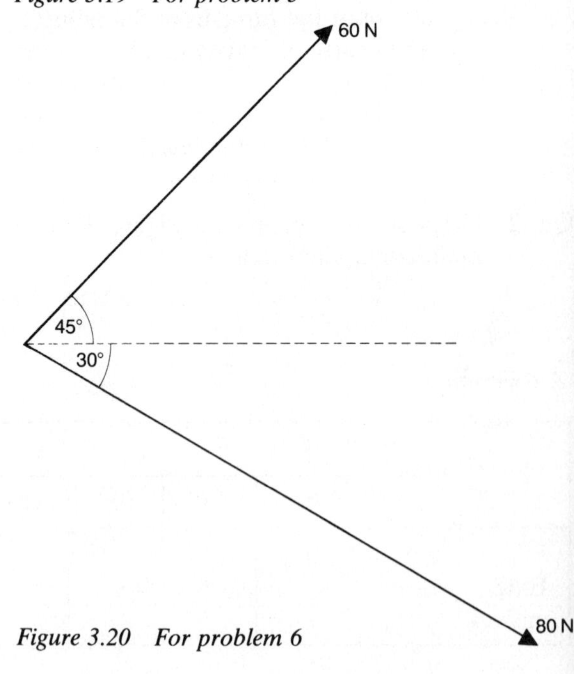

Figure 3.20 For problem 6

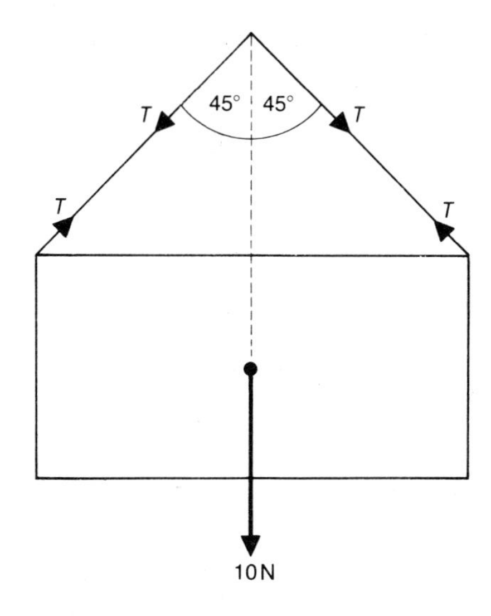

Figure 3.21 For problem 9

7 Determine, by constructing a vector diagram, the resultant of the two forces:

$F_1 = 100\,\text{N}\,\angle 0°$ and $F_2 = 75\,\text{N}\,\angle 90°$

acting at a common point.

8 By means of vector diagrams, determine the resultant of the three coplanar forces:

$F_1 = 100\,\text{N}\,\angle 0°$, $F_2 = 75\,\text{N}\,\angle 90°$, $F_3 = 80\,\text{N}\,\angle 120°$, acting at a common point.

9 A picture of weight 10 N is suspended by a cord as shown in Figure 3.21. Determine the tension force T in the cord.

4 Pressure in fluids

General learning objectives: to establish that fluid pressure at any level is equal in all directions, is normal to its containing surface, and is dependent on density and head of liquid.

4.1 The definition of pressure

Pressure is defined as normal force per unit area. By 'normal' we mean the force component acting at right-angles to unit area of the surface experiencing the pressure. This general definition describes quantitatively the force per unit area exerted on a surface by solids and fluids (fluid is the general term including both liquids and gases).

The solid block of material resting on the horizontal surface shown in Figure 4.1(a) exerts a vertical force on the surface due to gravitation attraction of the earth. If the mass of the block is m kilograms, this force (the weight) is mg newtons. If the area of the block resting on the surface is A square metres, then:

$$\text{Pressure } P = \frac{\text{normal force}}{\text{area over which force acts}}$$

$$= \frac{F}{A} = \frac{mg}{A}$$

newtons per square metre (N/m^2)

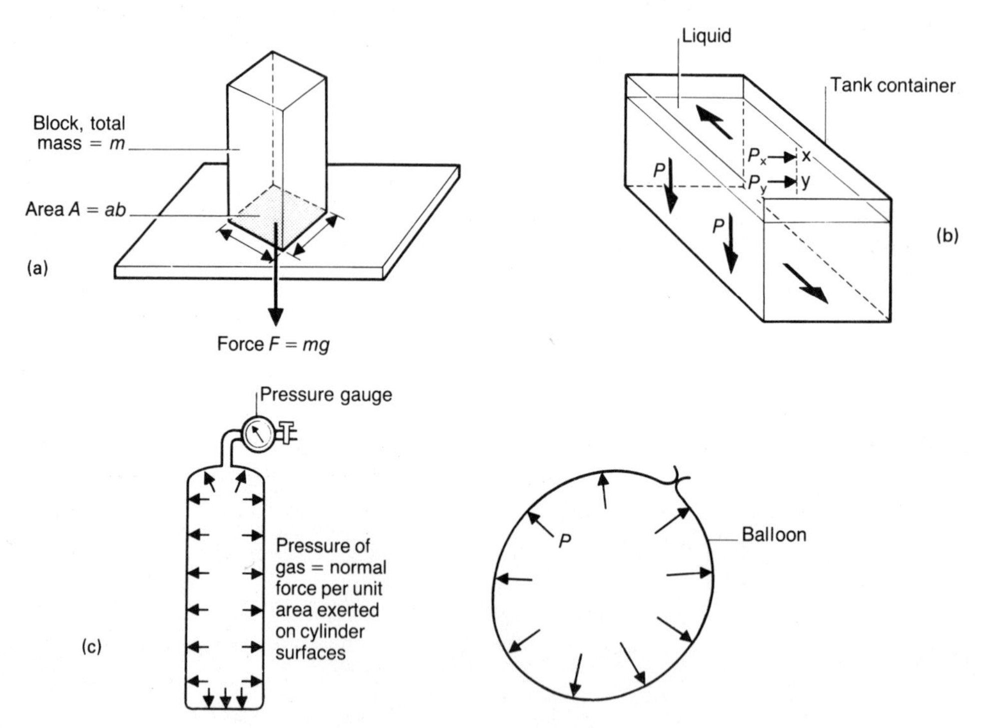

Figure 4.1
(a) Pressure on horizontal surface supporting block, $P = \frac{F}{A} = \frac{mg}{A}$ *N/m^2*
(b) Liquid in tank exerts pressure on its containing walls. Pressure acts at 90° to walls and increase with depth, i.e. $P > P_y > P_x$
(c) Pressure exerted by gas

A fluid in a container exerts forces on the container walls which are normal to these walls. For example the liquid in the tank shown in Figure 4.1(b) exerts a vertical force on the tank base and forces at right angles to the four vertical sides of the tank. If the total mass of liquid is m, the total force exerted on the base is $F = mg$ and the pressure over the base area, of area A, is

$$P = \frac{\text{normal force}}{\text{base area}} = \frac{F}{A} = \frac{mg}{A}\ \text{N/m}^2.$$

The forces on the vertical sides of the tank increase with depth and thus the pressure on these sides also increases with depth reaching a maximum at the tank bottom. The pressure P_x at point x on a vertical side equals the force per unit area at point x. P_x will be less than the pressure P_y at the deeper point y, shown in the diagram.

The gas contained in the cylinder shown in Figure 4.1(c) exerts forces normal to the cylinder walls. The force per unit area exerted on the walls is the gas pressure.

Note that pressure is 'transmitted' throughout a fluid and its effects are experienced at the surfaces of the container or at the surfaces of a body immersed in a fluid.

Since pressure is force per unit area, its units are newtons per square metre. The SI derived unit of pressure is the pascal (named after the famous French mathematician, Blaise Pascal), which has the abbreviation, Pa:

$1\ \text{Pa} = 1\ \text{N/m}^2$

Examples

1 (a) Determine the pressure exerted by a force of 5000 N acting over an area of 25 m^2.

(b) Determine the total force acting over an area of 0.25 m^2 experiencing a pressure of 200 Pa.

Solutions

(a) Pressure, $P = \frac{F}{A}$

$$= \frac{5000}{25} = 200\ \text{N/m}^2 \text{ or } 200\ \text{Pa}$$

Ans

(b) Force, $F = P \times A$

$= 200 \times 0.25 = 50\ \text{N}$ *Ans*

2 Determine the pressure on the base of a tank, base area $A = 2\ \text{m}^2$ and volume $v = 1.25\ \text{m}^3$ when filled with a liquid of density $\rho = 870\ \text{kg/m}^3$. Take $g = 9.81\ \text{m/s}^2$.

Solution

Mass of liquid, m = volume × density

$= v \times \rho = 1.25 \times 870$

$= 1087.5\ \text{kg}$

Therefore total force exerted by this mass on the tank base,

$$F = mg = 1087.5 \times 9.81 = 10.67 \times 10^3\ \text{kg}$$

hence the pressure on the base,

$$P = \frac{F}{A} = \frac{10.67 \times 10^3}{2} = 5.34 \times 10^3\ \text{N}$$

Ans

4.2 Pressure in fluids: properties and factors which determine pressure

We have defined pressure in general terms as force per unit area. In this section we discuss the

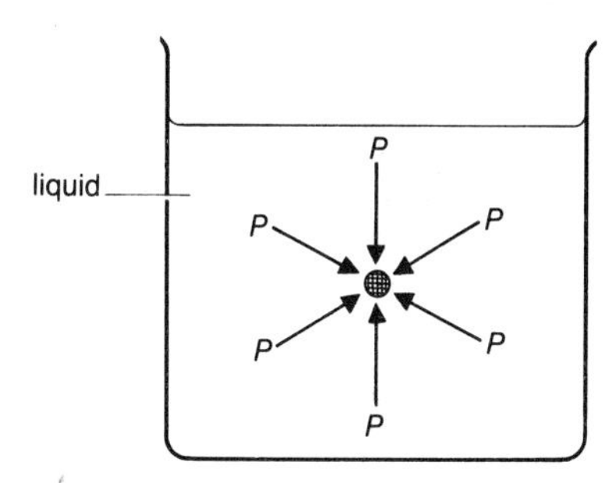

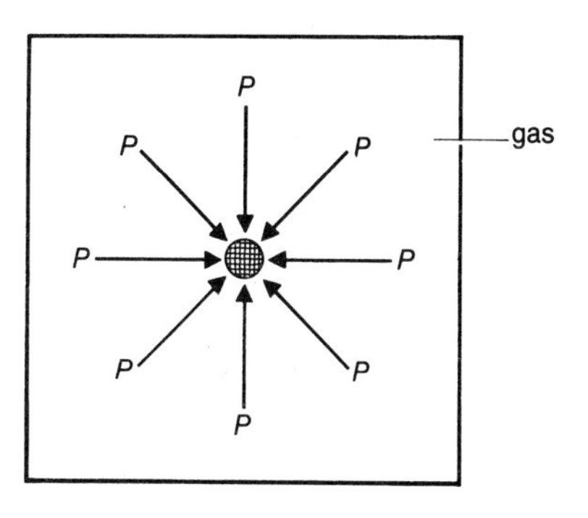

Figure 4.2 Pressure at any point in a fluid acts equally in all directions

properties and factors which determine the pressure in fluids.

In fluids – liquids and gases – pressure is transmitted in all directions throughout the fluid. If we were, for example, to place a tiny body in a liquid or gas as shown in Figure 4.2, the body would experience equal pressure in all directions. The pressure vectors act normally to the body's surface and are equal in magnitude as illustrated in the figures.

The pressures exerted by a fluid on the surfaces of its containing vessel are always normal (at 90°) to these surfaces, as illustrated in Figure 4.3. The pressure at any point in a fluid is also independent of the shape of the container. For example, in Figure 4.4, the three containers are of different shapes but, provided they are filled with the same liquid to identical heights, the pressures at any point at the same depth below the liquid surface in each container are equal.

The factors that determine the magnitude of pressure at any point in a fluid are:

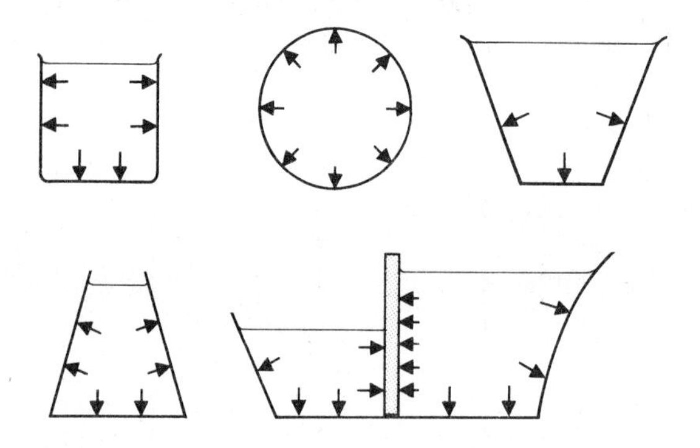

Figure 4.3 Pressure in a fluid acts in a direction normal (at 90°) to its containing surfaces

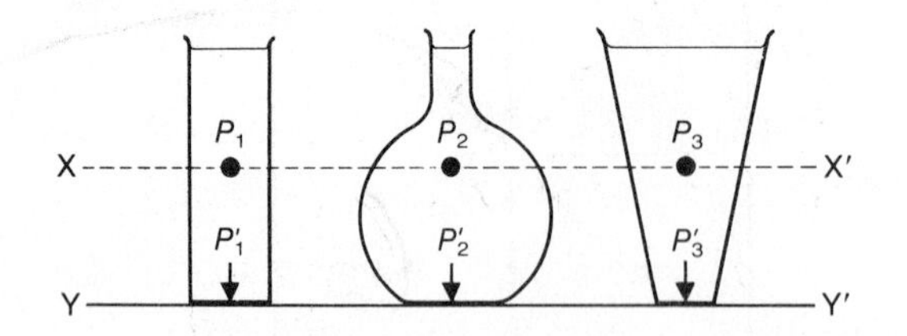

Figure 4.4 Pressure in a fluid at a given point in a fluid is independent of the shape of the container. e.g. at level XX': $P_1' = P_2' = P_3'$ at level YY': $P_1 = P_2 = P_3$

1 The density of the fluid

The greater the density of the fluid, the greater is the pressure. For example, the pressures at the same positions below the surface of the liquids in the three containers in Figure 4.5 are directly proportional to the density of the respective liquids; for example, if the first contains water, which has a density of $\rho_w = 1000\,kg/m^3$, the second turpentine of density $\rho_t = 870\,kg/m^3$ and the third mercury of density $\rho_m = 13\,600\,kg/m^3$, the pressures at equal depths below the surface are in the ratio:

$$\rho_w:\rho_t:\rho_m = 1:0.87:13.6$$

2 Head of fluid

The pressure is directly proportional to the head of fluid, i.e. the depth of the point considered below the surface of the fluid or equivalently the height of the fluid above the point. Thus, for example, in Figure 4.5 the pressure at points Q_2 will be double that at points Q_1 since the head of fluid is double.

3 *Surface pressure*

Pressure applied at the surface of a fluid is transmitted throughout the fluid and therefore will increase the pressure at any point in a liquid (gases in this sense will be compressed) by an amount equal to the pressure applied.

For example, in Figure 4.6, a surface pressure is applied to the liquid surface by means of a tightly fitting piston. If the force applied to the piston is F newtons and the area of the piston in contact with the liquid is A square metres, then the piston exerts a surface pressure of $P_s = F/A$ which is transmitted to all points in the fluid.

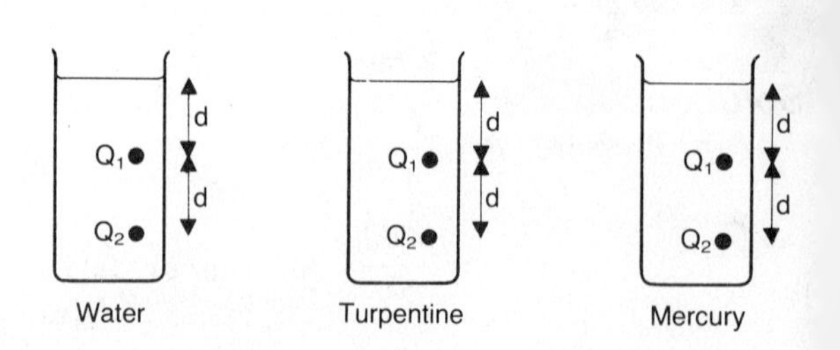

Figure 4.5 Pressure is directly proportional to fluid density and head of fluid

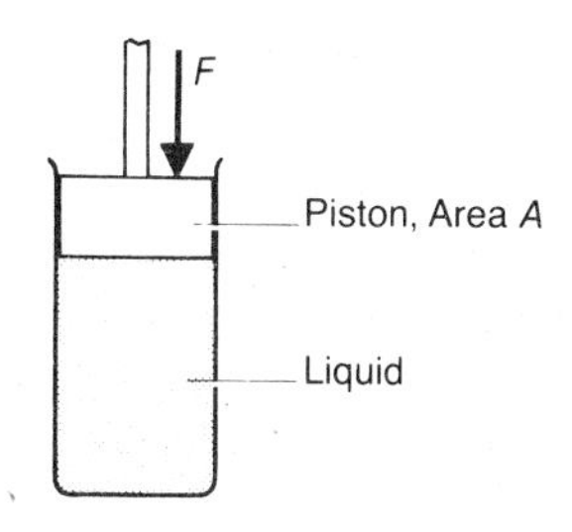

Figure 4.6 Surface pressure $P_s = F/A$ *is transmitted throughout liquid and the pressure at all points is increased by* P_s

Combining these three factors we can state that the pressure at any point in a fluid is given by

$$P = h\rho g \quad \text{N/m}^2 \text{ or Pa}$$

If surface pressure is P_s is also exerted, then.

$$P = h\rho g + P_s$$

where h = head of fluid, metres (m)
ρ = density of fluid, kg/m^3
$g = 9.81$ m/s^2, acceleration due to gravity.

We can derive these results by calculating the force exerted by the mass m of the fluid over a surface A at a depth h below the fluid surface (see Figure 4.7):

Total mass of fluid in head h,
m = volume of column × density
$= (A \times h) \times \rho = Ah\rho$ kilograms

so the gravitational force acting on area A is

$$F = mg = Ah\rho g$$

and the force per unit area, the pressure

$$P = F/A = h\rho g$$

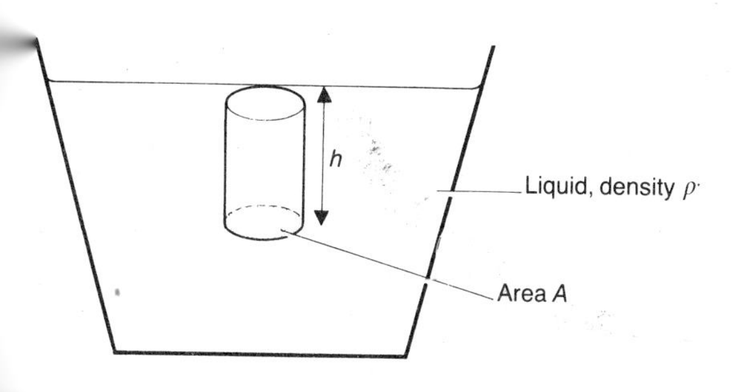

Figure 4.7 Derivation of pressure, $P = h\rho g$

If a surface pressure is also applied, P is increased by P_s.

Examples

1 Atmospheric pressure (the pressure due to the layers of the air in the earth's atmosphere) is often quoted as 760 mm mercury. This means that the atmospheric pressure is equal to the pressure exerted by a column of mercury of height 760 mm. Given that the density of mercury is $\rho_m = 13.6 \times 10^3$ kg/m^3, calculate the value of atmospheric pressure in pascals. Take $g = 9.81$ m/s^2.

Solution

Atmospheric pressure, $P_a = h\rho_m g$
where $h = 760 \text{ mm} = 0.76 \text{ m}$
$\rho_m = 13.6 \times 10^3 \text{ kg/m}^3$
so $P_a = 0.76 \times 13.6 \times 10^3 \times 9.81$
$= 1.014 \times 10^5$ Pa (or N/m^2) *Ans*

2 Calculate the pressure exerted on a submarine at a depth of 150 m below the surface of the sea. The density of sea-water is 1.025×10^3 kg/m^3 and $g = 9.81$ m/s^2.

Solution

Pressure at depth $h = 150$ m, due to the weight of sea-water acting on the submarine, is:

$P = h\rho g$
$= 150 \times 1.025 \times 10^3 \times 9.81$
$= 15.083 \times 10^5$ Pa *Ans*

Note: atmospheric pressure, $P_{atmos} \approx 10^5$ Pa, so the pressure on the submarine at this depth due to the weight of sea-water is approximately fifteen times atmospheric pressure. In addition the atmospheric pressure acts as a surface pressure, so the total pressure on the submarine at a depth of 150 m is 15 + 1 = 16 'atmospheres'.

4.3 Atmospheric pressure

The earth is surrounded by an atmosphere of air consisting mainly of the gases of nitrogen (78%) and oxygen (21%). The layers of air experience downward forces due to gravitational attraction and hence develop 'atmospheric' pressure exactly as pressure is developed by successive layers of

Table 4.1 Variation of atmospheric pressure with height above the earth

Height above sea-level (metres)	*Atmospheric pressure* (Pa)	(mm of mercury)
0	1.013×10^5	760
1000	0.90×10^5	674
2000	0.80×10^5	597
3000	0.70×10^5	525
4000	0.61×10^5	462
5000	0.54×10^5	407

liquid. The atmospheric pressure at any given height above the earth's surface is due to the weight of the air above us. We should therefore expect atmospheric pressure to be a maximum at sea-level and decrease with height above sea-level. This is indeed the case and the variation of pressure with height is given in Table 4.1.

The average pressure the atmosphere exerts at sea-level, measured at 0°C, is known as standard atmospheric pressure and has the value

$$101325 = 1.01325 \times 10^5\,\text{N/m}^2 \text{ or Pa}$$
$$= 1.01325 \text{ bar}$$

(1 bar $= 10^5\,\text{N/m}^2$ is still often used as a unit of pressure.)

Atmospheric pressure fluctuates about this value from day to day depending on temperature and weather conditions. A 'dropping' pressure normally indicates poor weather is on the way; a rapidly falling pressure means storms and gales.

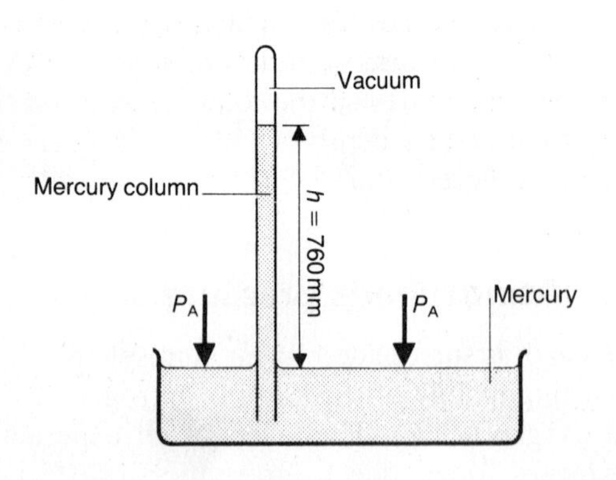

Figure 4.8 Atmospheric pressure (P) will support a column of mercury of height 760 mm

A rising pressure normally denotes a change to better weather.

The standard atmospheric pressure of 1.013×10^5 Pa will support a column of mercury 760 mm in height as shown in Figure 4.8 (see also example 1 in section 4.2). This provides a very convenient and accurate means of measuring atmospheric pressure in mercury barometers. A barometer is an instrument which measures atmospheric pressure.

Figure 4.9 shows an example of a mercury barometer, the Fortin barometer, which is a laboratory standard for atmospheric pressure measurement. The barometer tube, XZ, containing the mercury column is mounted vertically and is normally partially enclosed by a second metal tube for protection. At the base of the instrument, XZ dips into a reservoir of mercury. In order to obtain a constant zero reference level for the mercury column height measurements a fixed pointer is attached to the metal tube casing. The mercury level in the reservoir may then be raised

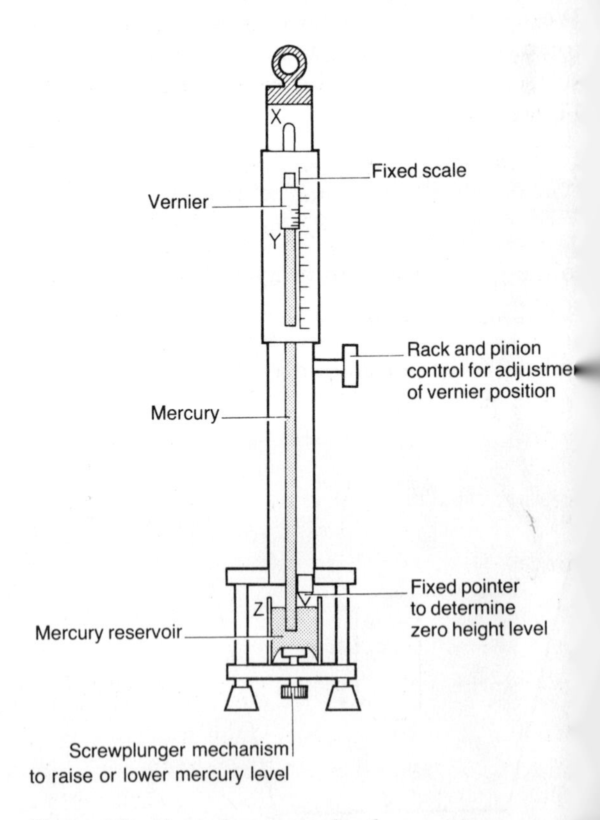

Figure 4.9 Fortin barometer for the precision measur ment of atmospheric pressure

or lowered by a screw-plunger mechanism, which pushes against the flexible base of the mercury reservoir container, until the level is seen to touch this pointer. The height of the mercury column is measured by means of the vernier and fixed scale, the vernier position being adjusted by a rack-and-pinion mechanism until the bottom of the vernier appears to touch the upper surface of the mercury column in the barometer tube at Y. A dummy vernier (not shown in the diagram) is provided at the back of the tube to avoid parallax errors.

A thermometer is normally provided, and is attached to the case of the barometer so that corrections may be made for temperature effects on the density of mercury and expansion or contraction of the scale length, which both effect the measured value of the mercury column height.

The long column of mercury essential to the Fortin barometer is impractical for many purposes. For less accurate requirements, and where automatic readings may be obtained without the necessity of adjustment, other more compact instruments have been developed. The best known of these is the Aneroid barometer which works basically on the same principle as the bellows pressure gauge described in section 4.5. In its simplest form it consists of a disc-shaped cylinder which has been pumped to a low pressure before being sealed. Its walls are made of thin flexible metal and are usually corrugated or pleated to increase both flexibility and surface area. Owing to the low interior pressure, the pressure on the external surfaces, i.e. atmospheric pressure, compresses the walls of the box. Any variation in atmospheric pressure thus causes a variation in this compression. These variations, which are normally very small, are communicated and magnified by a system of levers, and finally to pointer which registers atmospheric pressure directly on a calibrated scale.

.4 Absolute and gauge pressures

bsolute pressure is the pressure measured with respect to zero pressure. It is the total pressure including surface pressure such as atmospheric pressure.

auge pressure is the pressure measured by a pressure gauge in excess of atmospheric pressure. Thus

gauge pressure
= *absolute pressure minus atmospheric pressure*
and
absolute pressure
= *gauge pressure plus atmospheric pressure*

Most commercial–industrial pressure gauges register gauge pressure, the pressure in excess of atmospheric pressure. Note the absolute zero of pressure relates to that of a complete vacuum where no pressure whatsoever exists. On the other hand zero gauge pressure means that the pressure is equal to atmospheric pressure.

4.5 The measurement of pressure

4.5.1 The measurement of pressure using a U-tube manometer

The U-tube manometer is a simple instrument which can be used to measure pressure in both gases and liquids by measuring the difference in levels of a liquid in a U-tube.

Figure 4.10 illustrates the use of a manometer to measure gas pressure. The liquid in the U-tube is usually mercury, although for low-pressure measurements water can be used. In each of the diagrams one end of the U-tube is open to the atmosphere and hence the surface pressure acting upon the liquid column in this side of the tube is atmospheric pressure, which we will denote by P_A. The other end of the U-tube is connected to the gas, whose pressure we are to measure.

In (a) the levels of the liquid in the manometer are the same. Thus the pressure of the gas P in the container exerted on the left-hand column must exactly balance the atmospheric pressure P_A exerted on the right-hand column of liquid, thus

$$P = P_A \text{ N/m}^2$$

In (b) the pressure of the gas P_b must exceed atmospheric pressure, thus forcing the liquid in the U-tube round until the excess pressure caused by the height h of liquid causes the pressure on the two sides to equate. Hence the gas pressure in the container is

$$P_b = h\rho g + P_A \text{ N/m}^2$$

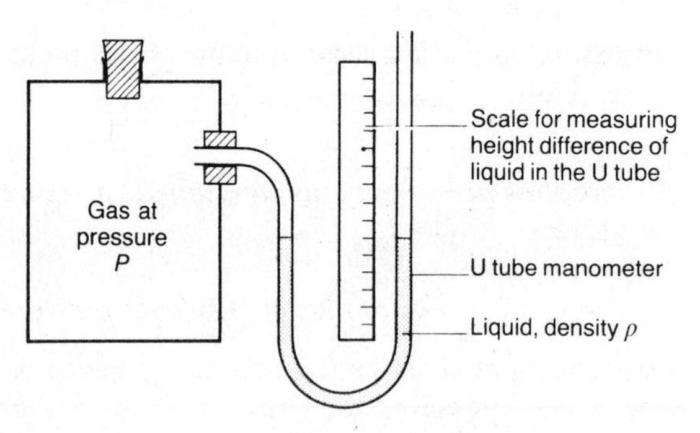

(a) Gas pressure P = atmospheric pressure P_A

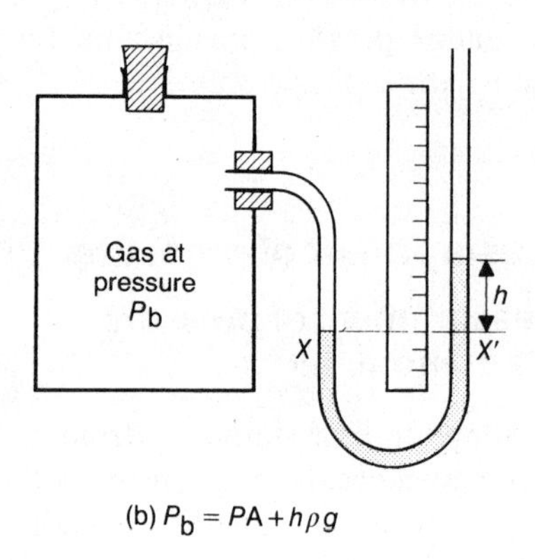

(b) $P_b = PA + h\rho g$

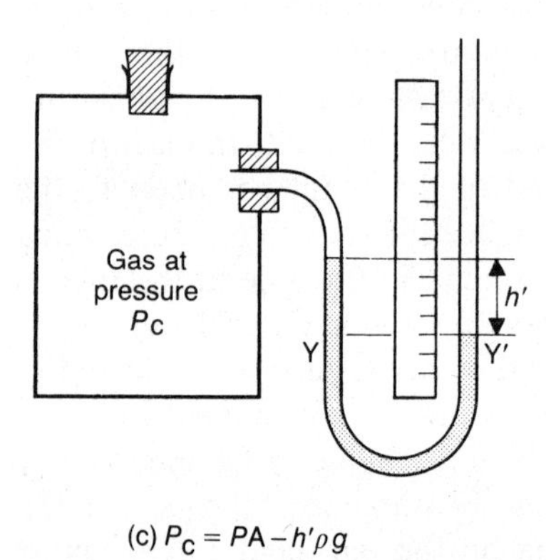

(c) $P_c = PA - h'\rho g$

Figure 4.10 The U-tube manometer: diagrams illustrating its use for measuring gas pressure

that is, atmospheric pressure plus the pressure due to the column of liquid $h\rho g$, where h = column height above XX′, ρ = liquid density, g = 9.81 m/s^2. Alternatively we could use the following argument: the pressure at X′ in the liquid is $h\rho g$ above atmospheric pressure and since pressures at the same level of a liquid are equal, the pressure at X′ equals the pressure at X, which in turn equals the gas pressure P_b in the container.

In (c) the gas pressure P_c in the container is below atmospheric pressure. The greater atmospheric pressure forces the liquid round in the U-tube until the excess pressure produced by the height h' of liquid causes the pressure on the two sides to equate. Hence the gas pressure in the container is

$$P_c = P_A - h'\rho g \text{ N/m}^2$$

Alternatively we may argue: pressure at Y′ = pressure at Y = atmospheric pressure, but pressure at Y is $h'\rho g$ greater than at point Q due to the column of liquid; hence pressure at Q is $P_A - h'\rho g$, which is the pressure of the gas in the container.

The open-end U-tube described above can be used to measure pressures up to about twice atmospheric pressure. When $P = 2P_A$, h = 760 mm, assuming mercury is used in the manometer. Thus higher pressures would require a length of U-tube in excess of 760 mm which is not very practical. The U-tube can be adapted to measure higher pressures by closing one end and calibrating the length of trapped air in this end in terms of pressure when the other end is attached to known pressures.

4.5.2 Measurement of gas pressure using pressure gauges

One of the most commonly used gauges employed to measure fluid pressure in industry is the Bourdon gauge, a sketch of which is shown in Figure 4.11. The pressure-sensing element consists of a flattened hollow metal tube, normally made from spring bronze or steel, which is bent into a circular shape. When the gauge is connected to measure gas or liquid pressure, the fluid enters the tube and the pressure it exerts tends to straighten the tube. The tube is fixed at the inlet side and the movement at the far end caused by the fluid pressure is communicated via a link to a rack-and-pinion mechanism which actuates a pointer. The pointer moves over a scale, calibrated to register pressure directly. For increased movement and therefore increased sensitivity the tube may be bent through several turns in the form of a spiral or helix.

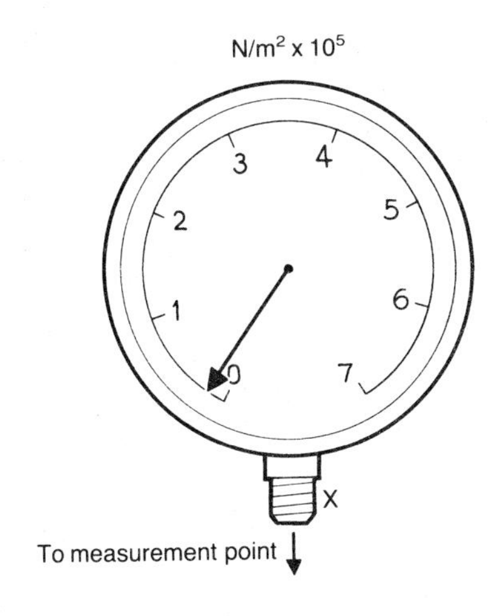

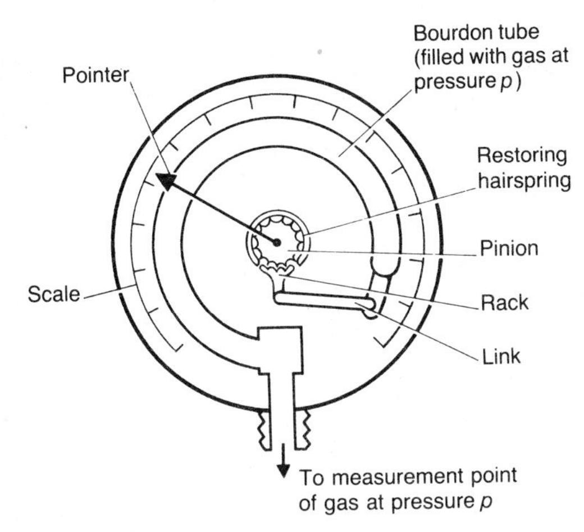

'igure 4.11 *A Bourdon pressure gauge*

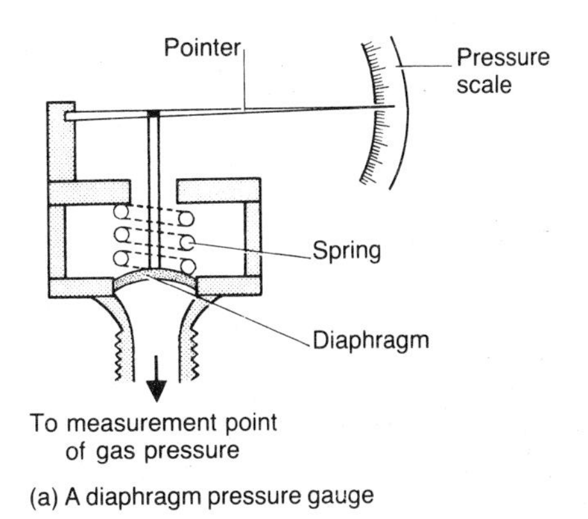

(a) A diaphragm pressure gauge

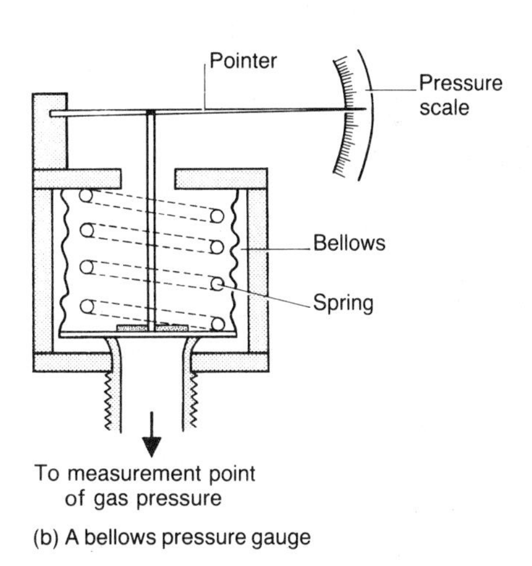

(b) A bellows pressure gauge

Figure 4.12 Diaphragm and bellows type pressure gauges

Figure 4.12 shows sketches of two other auges, which work on a similar principle. In the iaphragm gauge of (a) the fluid pressure acts on diaphragm which is thus caused to press against ıe spring shown. The displacement of the iaphragm, which depends on the fluid pressure ıd the balancing force of the spring, is communiıted via a rod to actuate a pointer which moves ver a scale calibrated to read pressure directly. he diaphragm may be made of metal, for cases here high strength and resistance to corrosion e wanted, or of rubber if high sensitivity is quired. The bellows-type gauge shown in (b) ıs the advantage of a greater range of deflection ver the diaphragm gauge. The pressure exerted the fluid at the base of the bellows is balanced by a spring and the displacement produced is transmitted via a rod to actuate a pointer.

There are many other types of pressure gauge available for a very wide variety of applications. At very low pressures encountered in high-vacuum work the McLeod gauge is frequently used. This gauge works on the principle of trapping a small amount of the gas whose pressure is required and then compressing this sample by a known amount and measuring the compressed sample pressure using a mercury U-tube manometer. The ratio of the actual to compressed sample gas pressure (i.e. the pressure measured by the manometer) is equal to the ratio of the compressed sample volume to the sample volume before compression. Using this technique, pressures of the order of a millionth or lower of atmospheric pressure may be measured.

4.6 Practical problems involving pressure

Examples

1 A specially designed deep-sea diving vessel can safely withstand pressures of up to $50\,\text{MN/m}^2$. Taking the mean density of sea-water as $\rho = 1050\,\text{kg/m}^3$ and $g = 9.81\,\text{m/s}^2$, determine the maximum depth at which the vessel could operate.

Solution

Let the maximum depth be h metres, then on equating the pressure $h\rho g$ at this depth to the maximum safe design pressure of $50 \times 10^6\,\text{N/m}^2$, we have

$$h\rho g = 50 \times 10^6$$

$$\text{so } h = \frac{50 \times 10^6}{1050 \times 9.81} = 4854\,\text{m}$$

Ans

2 In Figure 4.13 a force $F_1 = 50\,\text{N}$ is applied to the right-hand piston which has a surface area A. A weight of W newtons rests on the left-hand piston, which has an area of $10A$. The system is completely filled with water.

Neglecting the weight of the pistons, determine the value of W if the application of F_1 just causes it to be raised.

Solution

The application of F_1 produces a pressure $P = F/A$ which is transmitted through the water and causes an upthrust pressure on the

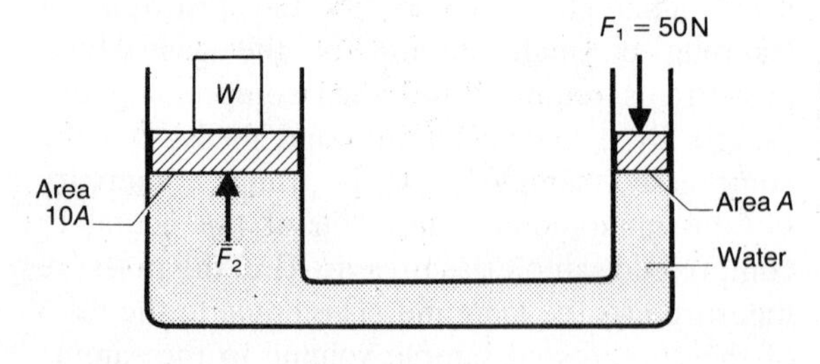

Figure 4.13 For example 2

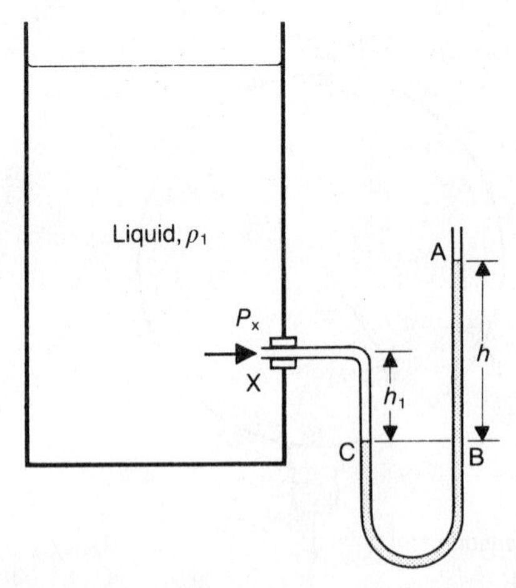

Figure 4.14 For example 3

left-hand piston carrying W. The resulting upward force P produces on this piston is

$$F_2 = P \times 10A$$

but as $P = F_1/A$,

$$F_2 = \frac{F_1}{A} \times 10A = 10F_1 = 10 \times 50 = 500\,\text{N}$$

Since this force is just sufficient to raise the piston carrying W, we have

$F_2 = W$, so $W = 500\,\text{N}$ *Ans*

3 Figure 4.14 shows a U-tube manometer used to measure the 'gauge' pressure P_X at point X in a tank of liquid of density ρ_1. The density of the liquid in the manometer is ρ_m. Show that

$$P_X = h\rho_m g - h_1\rho_1 g$$

If $\rho_1 = 1000\,\text{kg/m}^3$, $\rho_m = 13.6 \times 10^3\,\text{kg/m}^3$ and $h_1 = 0.1\,\text{m}$, $h = 0.25\,\text{m}$, determine the depth of point X below the liquid surface.

Solution

Consider the pressure of points B and C inside the manometer tube and at the same level, point C being at the interface between tank and manometer liquids.

The pressure at point B (in excess of atmospheric pressure) is

$$P_B = h\rho_m g$$

The pressure at point C,

$$P_C = P_X + \text{pressure due to column } h_1$$
$$= P_X + h_1\rho_1 g$$

Now since points B and C are at the same level, the pressures are equal, so

$$P_C = P_B$$
$$P_X + h_1\rho_1 g = h\rho_m g$$
$$\text{Thus } P_X = h\rho_m g - h_1\rho_1 g$$

If $\rho = 1.0 \times 10^3\,\text{kg/m}^3$, $\rho_m = 13.6 \times 10^3\,\text{kg/m}^3$, $h_1 = 0.1\,\text{m}$, $h_2 = 0.25\,\text{m}$, then:

$$P_X = 0.25 \times 13.6 \times 10^3 g - 0.1 \times 10^3 g$$
$$= (3.4 - 0.1)\,10^3 g = 3.3 \times 10^3 g$$

but $P_X = H\rho g = H \times 10^3 g$
where H = depth of point X, so

$$H = \frac{3.3 \times 10^3 g}{10^3 g} = 3.3\,\text{m} \qquad \textit{Ans}$$

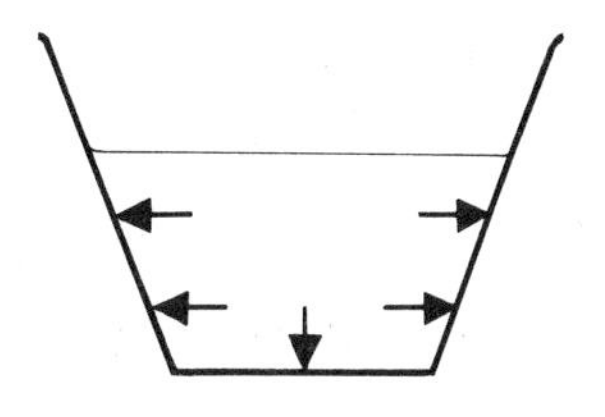

Figure 4.15 For Qu 1(c)

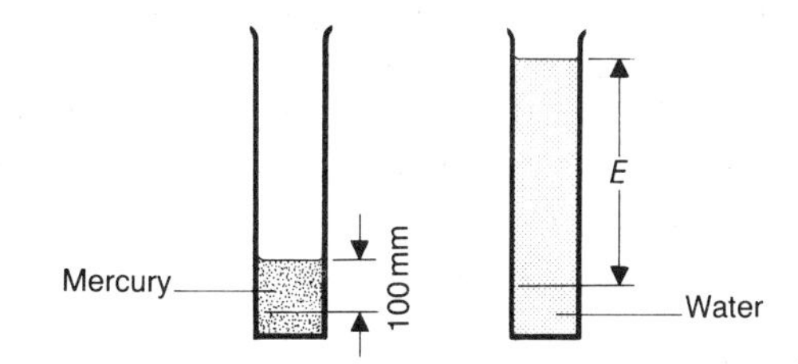

Figure 4.16 For Qu 1(d)

Test 4

This test may be used as a basic self-assessment test to check whether you have absorbed the main facts of Chapter 4 **Pressure in fluids** and its objectives. Enter your answers in the answer block.

Qu. 1 Enter a tick (√) in the answer block if you consider the statement correct; enter a cross (×) if you consider the statement in any way incorrect.

(a) Pressure is force per unit area and has the units of pascals: $1\,\text{Pa} = 1\,\text{N/m}^2$.

(b) The pressure at any given point in a column of liquid is greater in the vertical direction than in the horizontal direction.

(c) Pressure in a fluid always acts at right angles to the containing surfaces and hence the pressure directions marked in Figure 4.15 are all correct.

(d) The pressure at a depth 100 mm below the surface of a column of mercury is less than the pressure 1 m below a column of water, see Figure 4.16. The density of mercury is 13.6 times greater than that of water.

(e) Atmospheric pressure decreases in general with height above sea-level.

Qu. 2 Determine the pressure on the exterior frame of a submarine at a depth of 100 m below sea-level. The density of sea-water is $1025\,\text{kg/m}^3$, $g = 9.81\,\text{m/s}^2$ and atmospheric pressure is $1.013 \times 10^5\,\text{Pa}$.

Qu. 3 Determine the absolute pressure of the gas in containers (a), (b) and (c) in Figure 4.17. The liquid in the U-tubes is mercury of density $13.6 \times 10^3\,\text{kg/m}^3$. Take $g = 9.81\,\text{m/s}^2$ and atmospheric pressure is $1.013 \times 10^5\,\text{Pa}$.

Answer block:

Question no.	1					2	3		
	(a)	(b)	(c)	(d)	(e)		(a)	(b)	(c)
Answer									

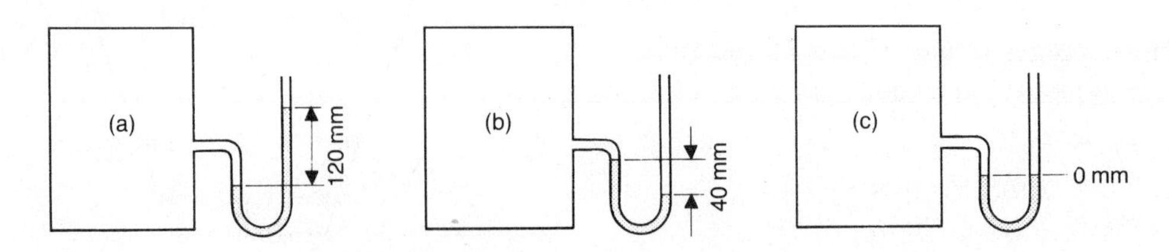

Figure 4.17 For Qu 3

Problems 4

1 Define pressure and state its units.
A rectangular tank has a base area of 1.5 m × 2 m and is filled to a height of 1.25 m with a liquid of density 900 kg/m^3. Calculate:
(a) the total force and pressure exerted on the tank base;
(b) the pressure on a tank side 0.6 m above the base. Take g = 9.81 m/s^2.

2 State the factors which determine the pressure at any point in a fluid.
Describe an experiment which demonstrates the relationship between pressure and head of liquid.

3 Distinguish between gauge and absolute pressure.
Describe how pressure may be measured using (a) a U-tube manometer, (b) a pressure gauge.

4 Draw and label a diagram of a precision barometer for atmospheric pressure measurement and explain its principle of operation.
The mercury column height measured on a Fortin barometer at a particular time is 748 mm of mercury. Convert this reading to pascals, given density of mercury $\rho_m = 13.6 \times 10^3$ kg/m^3 and g = 9.81 m/s^2.

5 A Bourdon pressure gauge has been calibrated in gauge pressure of pounds per square inch. Given 1 lb = 0.454 kg and 1 in^2 = 645 × 10^{-6} m^2, calculate the factor you should use to multiply the dial readings of the pressure gauge to convert them to newtons per square metre (N/m^2 or Pa). Take g = 9.81 m/s^2.

6 The boiling point of water depends on the pressure exerted on its free surface. Under standard atmospheric pressure (1.013 × 10^5 Pa or 760 mm mercury) water boils at 100°C. Given that a reduction of pressure causes the boiling point to fall at the rate of 3.6 × 10^3 Pa per degree Celsius for pressures close to standard atmospheric pressure, determine the boiling point of water when the pressure is (a) 748 mm of mercury; (b) 0.90 × 10^5 Pa.

Part Three: Motion and energy

5 Speed, velocity and distance–time graphs

General learning objectives: to obtain distance–time data, to plot distance–time graphs, to determine gradients of such graphs and to interpret slopes as speed, to find average speed and to explain why speed is a scalar quantity and velocity is a vector quantity.

5.1 The definition of speed and its calculation from given time–distance data

Speed is a measure of how quickly a body is moving. More formally, speed is defined as the ratio of the distance travelled to the time taken.

If the speed u is constant and the distance travelled in a time of t seconds is s metres, then

$$\text{speed} = \frac{\text{distance travelled}}{\text{time taken}},$$

$$u = \frac{s}{t} \text{ metres per second (m/s)}$$

$$\text{distance} = \text{speed} \times \text{time}, \; s = ut \text{ metres (m)}$$

$$\text{time taken} = \frac{\text{distance}}{\text{speed}}, \; t = \frac{s}{u} \text{ seconds (s)}$$

The SI unit of speed is the metre per second (m/s), although kilometre per hour (km/h) is often used:

$$1\,\text{m/s} = 3.6\,\text{km/h}$$

$$1\,\text{km/h} = \frac{1}{3.6} = \frac{5}{18} \text{ or } 0.2778\,\text{m/s}$$

If the speed varies with time, then we define the average or mean speed as

$$\text{average speed} = \frac{\text{total distance travelled}}{\text{total time taken}}$$

Examples

1 A train travelling at a constant speed starts a timer as it passes a given point and stops the timer when it reaches a point 480 m further up the line. The timer records 20.8 s. Calculate the speed of the train in m/s and km/h.

Solution

$$\text{Speed} = \frac{\text{distance}}{\text{time}} = \frac{480}{20.8} = 23.1\,\text{m/s} \quad \textit{Ans}$$

$$= 23.1 \times 3.6 = 83.2\,\text{km/h} \quad \textit{Ans}$$

as $1\,\text{m/s} = 3.6\,\text{km/h}$

2 Calculate the average speed of an athlete who runs 5000 m in 13 minutes 10 seconds.

Solution

$$\text{Time taken} = (13 \times 60) + 10 = 790\,\text{s}$$

$$\text{so average speed} = \frac{\text{distance covered}}{\text{time taken}}$$

$$= \frac{5000}{790} = 6.33\,\text{m/s} \quad \textit{Ans}$$

3 The average speed of an aircraft on a long-distance haul of 6400 km is 720 km/h. Calculate its flight time.

Solution

$$\text{Time taken} = \frac{\text{distance travelled}}{\text{average speed}}$$

$$= \frac{6400}{720} = 8.89\,\text{h} \quad \textit{Ans}$$

(0.89 h = 0.89 × 60 minutes = 53.4 min, so time can be expressed as 8 h 53 min).

5.2 Plotting distance–time graphs, determination of gradient, and interpretation of the gradient as speed

The distance–time data given in Table 5.1 was obtained for a sports car under test, with the car being driven under constant speed during successive time intervals. The time taken for the car to change its speed at the beginning of a new time interval was only a matter of seconds and was therefore neglected.

Table 5.1 Distance–time data for sports car test

	Time, t	*Distance, s*
A	0	0
B	30 min	60 km
C	1 h 30 min	125 km
D	2 h 45 min	300 km
E	4 h 0 min	350 km
F	5 h 0 min	450 km

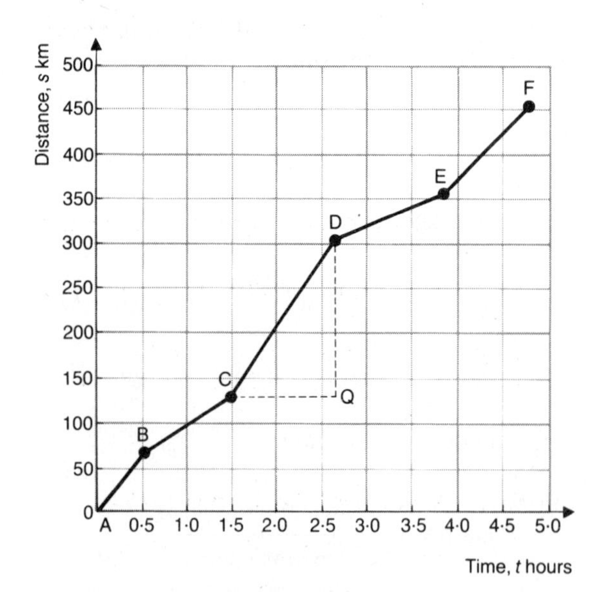

Figure 5.1 Distance–time graph for sports car test from data of table 5.1

The data in the table is represented in the distance–time graph shown in Figure 5.1. Distance in kilometres (1 unit = 50 km) is plotted vertically against time in hours (1 unit = 0.5 hour = 30 minutes) along the horizontal axis. The first point A ($t = 0$, $s = 0$) corresponds to the origin; the second point B ($t = 0.5$, $s = 60$) is plotted by moving 1 unit along the time axis (0.5 h) and then 1.2 units vertically, corresponding to a distance of 1.2 × 50 = 60 km. Likewise points C, D, E and F are plotted by first locating their t value horizontally and then moving vertically upwards the distance given by the s value.

Since the car travels with constant speed in each time interval, distance travelled increases linearly with time, we can join up the points A to B, B to C, C to D, D to E, E to F by straight lines.

The gradient or slope at any point on a distance–time graph, with distance plotted on the vertical axis and time on the horizontal axis, equals the speed at that point, i.e.

$$\begin{aligned}\text{speed} &= \text{rate of change of distance with time}\\ &= \text{gradient of distance–time graph}\end{aligned}$$

The gradient of a straight-line graph is a measure of its slope. Mathematically, gradient is defined as the ratio of the vertical distance between any two points on the straight line to the horizontal distance between the same two points.

Thus, referring to the distance–time graph of Figure 5.1 for our sports car test, we have, for example, over the time interval from C where $t = t_C = 1.5$ h to D where $t = t_D = 2.75$ h and where distance increases from $s = s_C = 125$ km to $s = s_D = 300$ km:

line CD defines the progress of distance with time;

$$\begin{aligned}\text{gradient of line CD} &= \frac{\text{vertical distance, DQ}}{\text{horizontal distance, QC}}\\ &= \frac{s_D - s_C}{t_D - t_C}\\ &= \frac{(300 - 125)\text{km}}{(2.75 - 1.5)\text{h}} = \frac{175}{1.25}\\ &= 140 \text{ km/h}\end{aligned}$$

which is, of course, equal to the speed of the car over this interval, as:

$$\text{speed} = \frac{\text{distance travelled}}{\text{time taken}}$$

$$= \frac{175\,\text{km}}{1.25\,\text{h}}$$

$$= 140\,\text{km/h}$$

The interpretation of gradient as equal to speed is especially important when the speed of a body is continually varying, as it allows us to determine the 'instantaneous' speed at any given point in time.

In Figure 5.2 the distance–time graph is plotted for a body whose speed is changing with time. Let us use this graph to determine the speed at three characteristic points P, Q and R marked on the graph.

1 At P, time $t = 2.2$ s

The gradient at point P is determined by first drawing in the tangent to the curve at P (i.e. place a ruler so it just touches the curve at P and draw in the line LM, the tangent). Next draw in the vertical line MN and horizontal line LN. Then:

$$\text{Gradient at P} = \frac{\text{MN}}{\text{NL}}$$

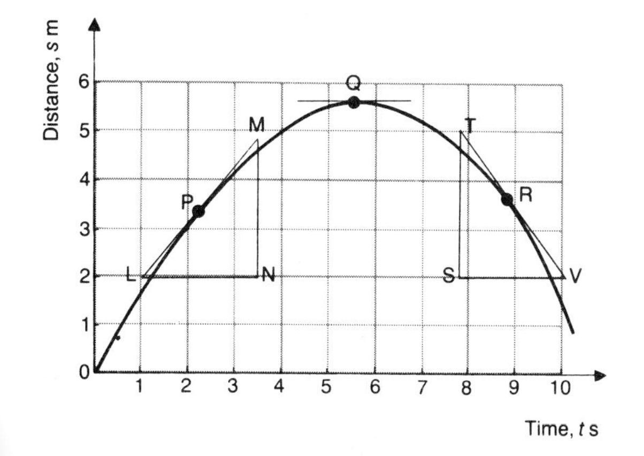

Figure 5.2 Distance–time graph for a body moving at variable speed: gradient at a given point = speed at that point

$$= \frac{5.0 - 2.0}{3.5 - 1.0} = \frac{3.0}{2.5} = 1.2\,\text{m/s}$$

= speed of body at P, when $t = 2.2$ s

2 At Q, time $t = 5.5$ s

Here the tangent is parallel to the time axis; there is for a very short time, about $t = 5.5$ s, no change in distance. The gradient and speed are therefore zero at this point.

3 At R, time $t = 8.8$ s

From point Q onwards we have the interesting case where distance is actually decreasing with time. This means the body must have changed the direction of its motion. From $t = 0$ to $t = 5.5$ s the body is travelling in a positive direction from a reference point, when $t > 5.5$ s the body is travelling in a direction back to its point of reference. The gradient at the point R is found by drawing the tangent to the curve at R:

$$\text{Gradient at R} = \frac{\text{ST}}{\text{VS}} = \frac{2.0 - 5.1}{10.1 - 7.8} = \frac{-3.1}{2.3}$$

$$= -1.35\,\text{m/s}$$

Note the gradient is negative as the distance travelled as measured with respect to the reference point is actually decreasing. The magnitude of the speed equals the magnitude of the gradient, 1.35 m/s. Including the negative sign provides extra information in defining the direction of motion.

Examples

1 Using distance–time graph of Figure 5.1, determine:

(a) the speed over the interval $t = 4$ to 5 h, E to F;

(b) the slowest speed used in the test;

(c) the average speed over the whole test.

Solution

(a) The speed over the interval E to F:

$$\text{speed} = \text{gradient of line EF}$$

$$= \frac{(450 - 350)\text{km}}{(5 - 4)\text{h}} = 100\,\text{km/h}$$

Ans

Table 5.2 Distance–time data for example 2

Time, *t* (seconds)	0	5	10	15	20	25	30	35	40	45	50
Distance, *s* (metres)	0	1.9	3.7	5.4	7.1	8.5	9.7	10.7	11.4	11.9	12.0

(b) The slowest speed corresponds to the region of smallest gradient. Clearly this is region DE, where

$$\text{speed} = \text{gradient DE}$$
$$= \frac{350 - 300}{4 - 2.75} = \frac{50}{1.25} = 40\,\text{km/h} \quad \textit{Ans}$$

(c) Average speed over whole test,

$$\bar{v} = \frac{\text{total distance}}{\text{total time}}$$
$$= \frac{450}{5} = 90\,\text{km/h} \quad \textit{Ans}$$

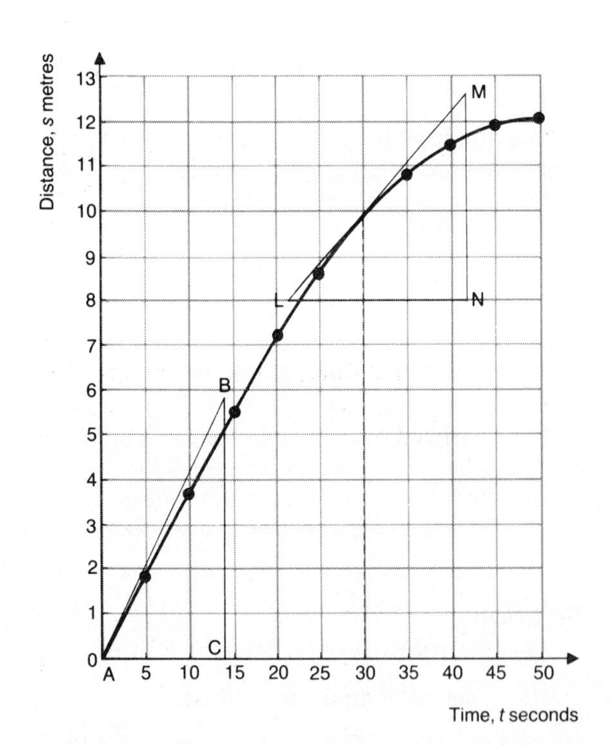

Figure 5.3 Distance–time for example 2 and determination of speed at t = *0 and* t = *30 s*

2 Table 5.2 gives distance–time data for a body whose speed varies with time. Plot the graph of distance versus time and use it to determine:
(a) the speed at time $t = 0$;
(b) the speed at time $t = 30$ s.

Solution

The distance–time curve is plotted in Figure 5.3 with a smooth curve drawn through the plotted points.

(a) Speed at time $t = 0$:

$$\text{Speed} = \text{gradient at } t = 0 = \frac{\text{BC}}{\text{AC}}$$
$$= \frac{6\,\text{m}}{14\,\text{s}} = \tfrac{2}{7} \text{ or } 0.43\,\text{m/s} \quad \textit{Ans}$$

(b) Speed at $t = 30$ s:

$$\text{Speed} = \text{gradient at } t = 30\,\text{s} = \frac{\text{MN}}{\text{LN}}$$
$$= \frac{12.7 - 8}{45 - 23} = \frac{4.7}{22} = 0.21\,\text{m/s} \quad \textit{Ans}$$

5.3 The distinction between speed and velocity

Speed = rate of change of distance with time
$$= \frac{\text{distance travelled}}{\text{time taken}}$$

Velocity = rate of change of distance with time, taking into account the direction of motion.

Speed is a scalar quantity. Speed defines the magnitude of the rate at which distance is covered and takes no account of the direction.

Velocity, however, contains in its definition the

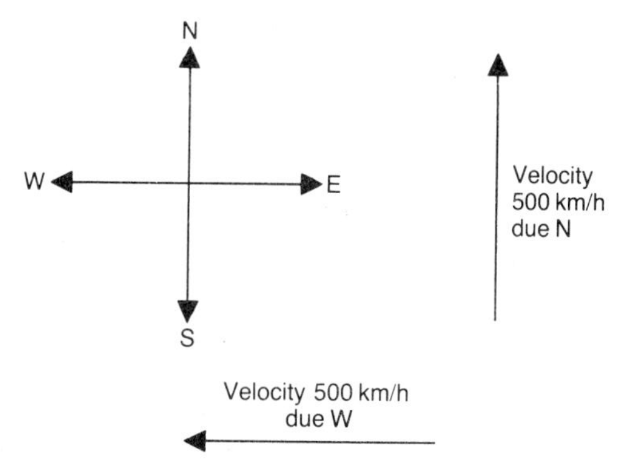

Figure 5.4 Diagrammatic representation of velocities of two aircraft flying at same speed but one in direction due N the other in direction due W (Scale 1 mm = 20 km/h)

speed and the direction of motion. Velocity is a vector quantity.

Speed and velocity are defined identically as far as magnitude is concerned and have the same units, but velocity also specifies the direction of motion. For example, an aircraft may be travelling due north at a speed of 500 km/h and another aircraft due west to 500 km/h. Their speeds are the same at 500 km/h, but their velocities would be defined, respectively, as:

500 km/h in direction due north
OR 500 $\angle 0°$ km/h on a bearing of 0° (i.e. due north).
500 km/h in direction due west
OR 500 $\angle 270°$ km/h on a bearing of 270° (i.e. 270° clockwise with respect to north).

Vector quantities, such as velocity, are represented diagrammatically by a line of length equal to the magnitude (suitably scaled) drawn in the direction the quantity acts, as explained in Chapter 3, section 3.2. Thus the velocities of the aircraft cited above could be represented as shown in Figure 5.4.

Test 5

This test may be used as a basic self-assessment test to check whether you have absorbed the main facts of the fifth chapter on **Speed, velocity and distance–time graphs**. All answers to be entered in the answer block.

Qu. 1 Enter a tick (√) in the answer block if you consider the statement correct; enter a cross (×) if you consider the statement in any way incorrect.

(a) Speed is the rate of change of distance with time.
(b) Average speed equals total distance travelled divided by time taken.
(c) The gradient of a distance–time graph at a given point equals the instantaneous speed at that point.
(d) Velocity differs from speed in that it defines both the magnitude in m/s and direction of the motion.
(e) The earth rotates about the sun in an approximately circular orbit at a constant velocity.

Qu. 2 (a) A particle travelling at constant speed in a straight line passes two points 10 m apart at successive times of 2.4 s and 6.8 s. Calculate the speed of the particle.
(b) Determine the average speed of an athlete who runs 1500 m in 3 min 45 s.
(c) A hovercraft travelling at an average speed of 70 km/h completes a cross-channel journey in 35 minutes. Determine the total distance travelled.
(d) Given that light travels at 3×10^8 m/s and that the distance between the sun and earth is 150×10^9 m, estimate the time taken for light leaving the sun to reach the earth.

Answer block:

Question no.	1					2				3		
	(a)	(b)	(c)	(d)	(e)	(a)	(b)	(c)	(d)	(a)	(b)	(c)
Answer												

Qu. 3 The distance–time data for a body travelling at varying speeds are given in the table below.

Time (s)	0	1	2	3	4	5	6
Distance (m)	0	13.5	19	23	25.5	27.5	29

(a) Determine the average speed over the 6 s period. Plot the distance–time graph and determine:
(b) the distance travelled at $t = 3.4$ s;
(c) the gradient and hence speed at $t = 2$ s.

Problems 5

1 (a) Calculate the average speed of a body which travels 100 m in 5 s.
(b) Calculate the average speed of an athlete who runs 10 000 m in 28 minutes and 2 seconds.

2 (a) Calculate the time taken for sound to travel 500 m if the speed of sound is 331 m/s.
(b) Calculate the time taken for a radio wave to travel from the earth to the moon. The earth–moon distance is 384,000 km and radio waves travel at the speed of light, 3×10^8 m/s.

3 The distance–time data for a body travelling at a varying speed are given in the following table.

Time, t (s)	0	5	10	15	20	25
Distance, s (m)	0	5	7.5	9.5	11	12

Determine:
(a) the average speed in the first 10 s;
(b) by plotting the distance–time,
(i) the speed when $t = 5$ s;
(ii) the time taken to travel 10 m.

6 Acceleration, velocity–time graphs and solving problems in linear motion

General learning objectives: to construct velocity–time graphs from given data, to calculate the gradient and to interpret the gradient as acceleration and to solve simple problems using graphical and calculation techniques.

6.1 The definition of acceleration

In the last chapter we defined velocity as the rate of change of distance with time with direction of the motion also specified. **Acceleration** is the *rate of change of velocity with time*. Acceleration has the units of metres per second per second or metres per second squared, m/s^2. Acceleration, like velocity, is a vector quantity and must be specified by giving both its magnitude and direction.

If the velocity of a body travelling in a straight line at a time t_1 seconds is v_1 metres per second and is experiencing constant acceleration such that its velocity at a later time t_2 s is v_2 m/s, then:

$$\text{Acceleration, } a = \frac{\text{change in velocity}}{\text{time taken}}$$

$$= \frac{v_2 - v_1}{t_2 - t_1} \text{ m/s}^2$$

If v_2 is greater than v_1, a is a positive acceleration; if v_2 is less than v_1 then a is a negative acceleration, also known as deceleration. When velocity changes at a constant rate, the acceleration is constant. If the rate of change is not constant, then a defined above is the average value of the acceleration between t_1 and t_2.

For bodies experiencing constant acceleration in a straight line we can establish a very useful formula:

$$v = u + at$$

where v = final velocity or velocity after t seconds

u = initial velocity, velocity at $t = 0$

a = acceleration

Examples

1 Calculate the acceleration of a body moving in a straight line and subjected to a constant force which produces a constant rate of change of velocity such that:

at $t = t_1 = 4$ s, velocity $v = v_1 = 10$ m/s
at $t = t_2 = 9$ s, $v = v_2 = 30$ m/s

Solution

$$\text{Acceleration } a = \frac{\text{change in velocity}}{\text{time taken}}$$

$$= \frac{v_2 - v_1}{t_2 - t_1} = \frac{30 - 10}{9 - 4}$$

$$= \frac{20}{5} = 4 \text{ m/s}^2 \quad \textit{Ans}$$

2 A car has a speed of 20 m/s at the instant its brakes are applied and a speed of 5 m/s when its brakes have been in operation for 4 s. The motion of the car is in a straight line. Calculate the average value of acceleration during the braking.

Solution

Average acceleration

$$= \frac{\text{final velocity minus initial velocity}}{\text{time brakes are applied}}$$

$$= \frac{5 - 20}{4} = \frac{-15}{4} = -3.75 \text{ m/s}^2 \quad \textit{Ans}$$

Note: the negative sign indicates a deceleration, the velocity decreases with increasing time.

3 A particle with an initial velocity of 4 m/s is accelerated by a constant force producing an

acceleration of $3\,m/s^2$. Determine its velocity after 8 s.

Solution

Using the formula,

$$v = u + at$$

with $u = 4\,m/s$, $a = 3\,m/s^2$ and $t = 8\,s$, we have:

$v = 4 + (3 \times 8) = 28\,m/s$ *Ans*

6.2 Plotting of velocity–time graphs for linear motion and interpretation of gradient as acceleration

In this section we plot velocity–time graphs for bodies moving in straight lines – linear motion. Examples of linear motion include cars travelling in straight lines, trains travelling down straight tracks, bodies falling under gravity, a spring oscillating in the direction of its axis, and the swings of a simple pendulum (provided the oscillations are relatively small).

Velocity–time graphs are important since they not only provide v–t information in graphical form showing how velocity varies with time but the graphs may be also used to find

(i) the acceleration at any point:

gradient of v–t graph = acceleration at point

(ii) the total distance s travelled:

$$s = \text{area under } v\text{–}t \text{ curve}$$

Examples

1 The velocity–time data given in Table 6.1 was measured for a body moving with uniform acceleration in a straight line.

Plot the velocity–time graph and determine the acceleration and the total distance travelled over the period 0 to 5 s

Table 6.1

Time, t (s)	0	1	2	3	4	5
Velocity, v (m/s)	0	6.5	13.2	19.8	26.5	33.0

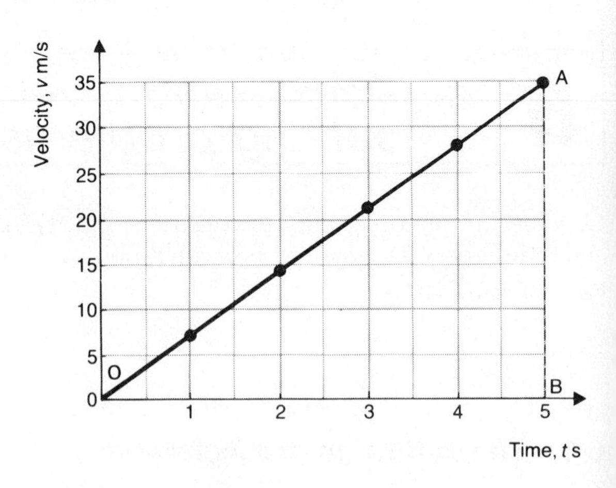

Figure 6.1 Velocity–time graph for example 1

Solution

The velocity–time graph is plotted in Figure 6.1. The points to a very good degree of approximation lie on the straight line, showing that the velocity increases linearly with time.

$$\text{Acceleration, } a = \frac{\text{change in velocity}}{\text{time taken}}$$

= gradient of straight-line graph

$$= \frac{33}{5} = 6.6\,m/s^2 \quad \textit{Ans}$$

Since the acceleration is constant, the average value of the velocity is:

$$v = \tfrac{1}{2}(\text{initial velocity} + \text{final velocity})$$
$$= \tfrac{1}{2}(0 + 33) = 16.5\,m/s$$

Hence, since:

distance travelled = average velocity × time

then the distance travelled in the 5 s is

$16.5 \times 5 = 82.5\,m$ *Ans*

The distance travelled can also be calculated by finding the area under the velocity–time graph distance travelled = area under velocity–time graph

$= \text{area } \triangle \text{ OAB}$

$= \frac{1}{2} \text{ base} \times \text{height}$

$= \frac{1}{2} \times 5 \times 33 = 82.5\,\text{m}^2$

2 Plot the velocity–time graph for the data given in Table 6.2 and determine the acceleration over the periods:

(i) $t \geqslant 0$ to $t = 20\,\text{s}$; (ii) $t > 20$ to $t = 50\,\text{s}$

Calculate also the total distance travelled in the 50 s.

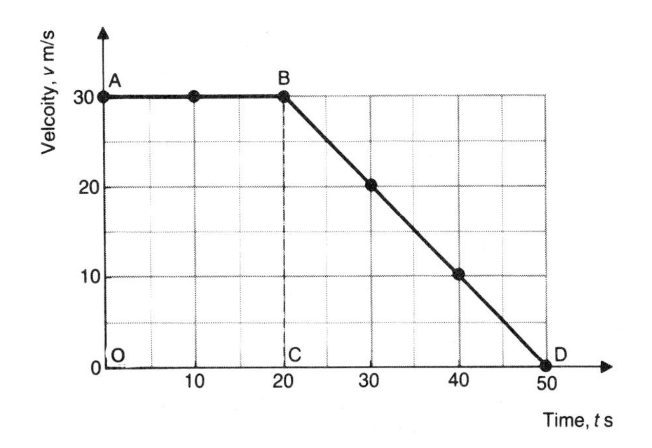

Figure 6.2 Velocity–time graph for example 2

Table 6.2

Time t (s)	0	10	20	30	40	50
Velocity v (m/s)	30	30	30	20	10	0

Solution

The velocity–time graph is plotted in Figure 6.2.

(i) Over the period $t \geqslant 0$ to 20 s, the velocity remains constant and hence the acceleration is zero:

$a = 0$ *Ans*

(ii) Over the period $t \geqslant 20$ to 50 s, the velocity decreases at a constant rate and the acceleration,

$$a = \frac{\text{change in velocity}}{\text{time taken}}$$

$$= \frac{0 - 30}{50 - 20} = \frac{-30}{30} = -1\,\text{m/s } \textit{Ans}$$

The total distance travelled,

s = area under velocity–time graph

= area OABC + area $\triangle$ CBD

$= (30 \times 20) + (\frac{1}{2} \times 30 \times 30)$

$= 1050\,\text{m}$ *Ans*

3 The velocity–time data of Table 6.3 was obtained for oscillatory motion (see inset in Figure 6.3). Plot the velocity–time graph and use it to determine the acceleration at

(i) $t = 0.15\,\text{s}$; (ii) $t = 0.5\,\text{s}$

Estimate also the distance travelled over the period $t = 0.75$ to 1 s

Solution

The velocity–time graph is plotted in Figure 6.3 and a smooth curve drawn through the points.

(i) Acceleration at $t = 0.15\,\text{s}$ corresponds to the gradient at this point, i.e.

acceleration at $t = 0.5\,\text{s}$ = gradient at P (see graph)

$$= \frac{\text{AB}}{\text{BC}} = \frac{0.6 - 11}{0.25 - 0.05} = \frac{-10.4}{0.2}$$

$= -52\,\text{m/s}^2$ *Ans*

(ii) Acceleration at $t = 0.5\,\text{s}$ is zero, since the

Table 6.3

Time, t (s)	0	0.1	0.2	0.3	0.4	0.5	0.6	0.7	0.8	0.9	1.0
Velocity, v (m/s)	10	8.1	3.1	−3.1	−8.1	−10	−8.1	−3.1	3.1	8.1	10

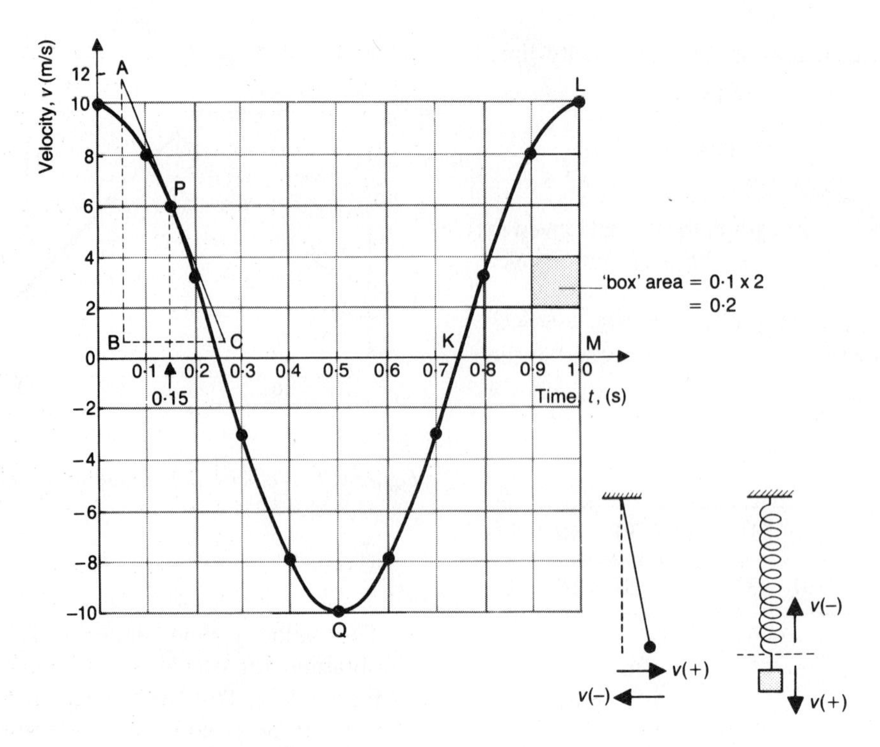

Figure 6.3 Velocity–time graph for oscillitory motion, see example 3

tangent at this point, point Q on Figure 6.3, is parallel to the time axis and hence the gradient is zero. At $t = 0.5$ s the rate of change of velocity and hence the acceleration is zero.

The distance travelled over the period $t = 0.75$ s to $t = 1$ s,

s = area under v–t curve from $t = 0.75$ s to $t = 1$ s.

To find this area we can count the number of 'squares' under the curve, taking the 'box' shown shaded in Figure 6.3 as our unit square:

Area of box = 0.1 s × 2 m/s = 0.2 m
Number of boxes under curve (approximating for fractional boxes)
= 5 whole boxes + 3 (equivalent from fractional boxes)
= 8
so area under curve = 8 boxes = 8 × 0.2 = 1.6 m
Hence $s = 1.6$ m *Ans*

6.3 Force and the relationship between force and acceleration

A force may be taken as a measure of the effect which is capable of producing and/or changing the course of motion. In Chapter 2 we described the effect of force in extending, compressing and shearing materials. In Chapter 3 we considered the conditions for forces and moments to establish equilibrium so that no motion was produced. Here we consider the relationship between force and the acceleration it produces on a body.

Based on Newton's second law of motion, we define the force F acting on a body of mass m in terms of this mass and the acceleration produced by the application of the force, as

$$F = m \times a \text{ newtons (N)}$$

where m = mass of body in kilograms (kg)
a = acceleration of body in m/s^2

$$\text{Thus } a = \frac{F}{m} \text{ m/s}^2$$

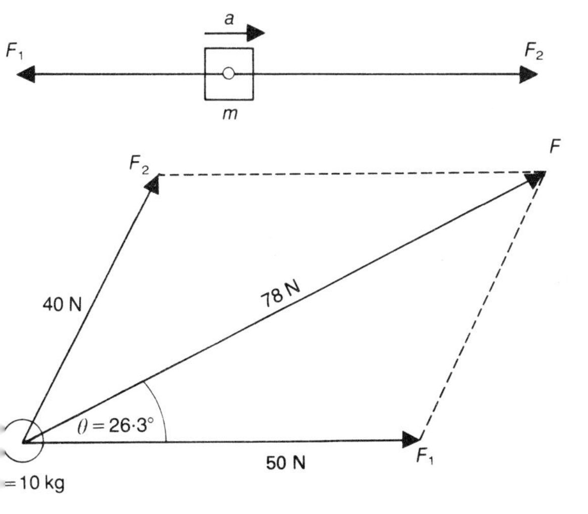

'igure 6.4 *Acceleration = Resultant force/mass*
a) Acceleration a $= \frac{F_2 - F_1}{m}$
b) Acceleration a = *7.8 m/s² ∠26.3°*

'he acceleration of a body is therefore directly roportional to the resultant force acting on the ody. Hence, for example, in Figure 6.4(a), where a ody of m kilograms is acted upon by two forces vhose lines of action are in opposite directions, the esultant force is $F_2 - F_1$, assuming F_2 is greater han F_1, and the acceleration produced by the esultant force on m is

$$a = \frac{F_2 - F_1}{m}$$

n Figure 6.4(b) the resultant force F must first be ound using the parallelogram of forces method (see ection 3.5), then the magnitude of a:

$$|a| = \frac{|F|}{m} = \frac{78\,\text{N}}{10\,\text{kg}} = 7.8\,\text{m/s}^2$$

Note: By convention the lines | | indicate the nagnitude of a quantity)
nd the direction of a and the resulting motion is

$$\angle\theta = 26.3°$$

.4 Gravitational force

Vhen a body is 'free-falling' under gravity, e.g. a ody dropped from a height above the earth's urface, the resultant force acting on the body is the gravitational force of attraction between the body and the earth minus the force of resistance due to the body's motion through the air and a small buoyancy force due to the body displacing a volume of air (Archimedes' force).

If we neglect the latter two forces, the gravitational force alone produces an acceleration of $g = 9.81\,\text{m/s}^2$ and if the mass of the body is m kilograms, the gravitation force is

$$F = mg \text{ newtons}$$

The numerical value of g actually varies slightly from point to point on the earth's surface, a maximum of ±0.4% about its average value. The International Standard value of g is $9.80665\,\text{m/s}^2$ as measured at 45° latitude and sea-level.

In practice, a body undergoing 'free-fall' first accelerates with an acceleration close to g. As its velocity increases the air resistance increases and so the net force on the body and hence its acceleration decreases. The body eventually reaches a terminal velocity where the gravitational force of attraction is exactly counterbalanced by resistive forces and, to a very small extent, buoyancy forces. The terminal velocity of medium size raindrops is about 8 m/s; the terminal velocity of a parachutist in a delayed-opening (free-fall) jump may be as high as 70 m/s (250 km/h) although his/her landing speed with parachute now open is about 5 to 10 m/s (20 to 30 km/h).

Examples

1 A hammer is dropped from the top of a radio mast and takes 3.5 s to reach the ground. Neglecting resistive forces, etc., and taking $g = 9.8\,\text{m/s}^2$, calculate:
(a) the velocity at which the hammer hits the ground;
(b) the average velocity in its fall;
(c) the height of the radio mast.

Solution

(a) If we assume the hammer is dropped with zero initial velocity, then the final velocity at time $t = 3.5$ s when it hits the ground is:

$$v = at = gt = 9.8 \times 3.5 = 34.3\,\text{m/s}$$

Ans

(b) Since the hammer experiences constant acceleration, its average velocity during the fall is

$$\bar{v} = \tfrac{1}{2}(\text{initial} + \text{final velocity})$$
$$= \tfrac{1}{2}(0 + 34.3) = 17.15\,\text{m/s}$$

Ans

(c) The height of the mast equals the distance travelled in the fall, so:

$$\text{mast height} = \text{average velocity} \times \text{time of fall}$$
$$= 17.15 \times 3.5 = 60.0\,\text{m}$$

Ans

2 A vertical force of $F = 300\,\text{N}$ is applied to lift a mass of 25 kg. Determine the resultant force on the mass and its acceleration. Take $g = 9.81\,\text{m/s}^2$.

Solution

Resultant force = applied force minus gravitational force

$$F_R = F - mg$$
$$= 300 - (25 \times 9.81) = 54.75$$

Ans

Also, since

$$F_R = ma$$

the acceleration,

$$a = \frac{F_R}{m} = \frac{54.75}{25} = 2.19\,\text{m/s}^2$$

Ans

3 A stone is projected vertically upwards with a velocity of 20 m/s. Estimate the maximum height it reaches and the total time taken to return to earth. Take $g \approx 10\,\text{m/s}^2$.

Solution

Let t = time to reach maximum height. At the maximum height the velocity of the stone must be zero (otherwise it would travel further upwards), so on applying the general formula

$$v = u + at$$

with v = final velocity = 0
u = initial velocity = 20 m/s
a = acceleration = $-g = -10\,\text{m/s}^2$ (the minus sign is included since gravity opposes motion on the upwa[rd] path),

we have $0 = 20 - 10t$
so $t = 20/10 = 2\,\text{s}$

The average velocity on the upward path,

$$\bar{v} = \tfrac{1}{2}(\text{initial} + \text{final velocity})$$
$$\bar{v} = \tfrac{1}{2}(20 + 0) = 10\,\text{m/s}$$

so maximum height reached,

$$\text{height} = \bar{v} \times t = 10 \times 2$$
$$= 20\,\text{m} \quad \textit{Ans}$$

The total time taken is twice that taken [to] reach the maximum height,
i.e. $2t = 2 \times 2 = 4\,\text{s}$ *Ans*

On the upward path gravity opposes moti[on] and slows velocity from 20 m/s to 0; on t[he] downward path gravity accelerates the sto[ne] from 0 to 20 m/s, neglecting any resisti[ve] forces, and hence the duration times for b[oth] paths are equal.

6.5 Solution of problems relating t[o] linear motion using formulae

For linear motion with constant acceleration [the] following important formulae may be used:

$$v = u + at$$
$$s = \text{average velocity} \times t \quad (2)$$
$$s = ut + \tfrac{1}{2}at^2 \quad (3)$$
$$v^2 = u^2 + 2as$$

where v = final velocity, m/s
u = initial velocity, m/s
t = time, s
s = distance, m
a = acceleration, m/s²

Formula (1) is derived directly from [the] definition of acceleration,

$$a = \frac{v - u}{t}$$

so $at = v - u$ or $v = u + at$

For constant acceleration a, the avera[ge] velocity

$$\bar{v} = \tfrac{1}{2}(\text{initial velocity} + \text{final velocity})$$
$$= \tfrac{1}{2}[u + (u + at)] = u + \tfrac{1}{2}at$$

and since distance travelled,

$$s = \bar{v} \times t$$

we have, on substituting for $\bar{v}$,

$$s = t(u + \tfrac{1}{2}at) = ut + \tfrac{1}{2}at^2$$

which establishes formula (2b).

It is often useful to eliminate t and obtain a formula linking u, v, a and s. Thus if we use (1),

$$t = \frac{v - u}{a}$$

and substituting this value for t in (2b) we have,

$$s = u\left(\frac{v - u}{a}\right) + \tfrac{1}{2}a\left(\frac{v - u}{a}\right)^2$$

On simplifying by multiplying throughout by $2a$, we obtain

$$\begin{aligned} 2as &= 2u(v - u) + (v - u)^2 \\ &= 2uv - 2u^2 + v^2 - 2vu + u^2 \\ &= v^2 - u^2 \\ \text{so } v^2 &= u^2 + 2as \end{aligned}$$

Let us now apply these formula to solve some practical problems.

Examples

1 A car starting from rest accelerates at a constant rate and reaches a velocity of 20 m/s in 10 s. Calculate the acceleration and the distance travelled.
If the brakes are then applied and produce a constant deceleration of 4 m/s^2, calculate the time taken for the car to come to rest, and the braking distance.

Solution

$$\text{Acceleration } a = \frac{20 - 0}{10} = 2\,\text{m/s}^2 \quad \textit{Ans}$$

The distance travelled can be found directly from formula (2b), i.e. $s = ut + \frac{1}{2}at^2$, by substituting $u = 0$, $a = 2\,\text{m/s}^2$ and $t = 10\,\text{s}$:

$$s = 0 + \tfrac{1}{2} \times 2 \times (10)^2 = 100\,\text{m} \quad \textit{Ans}$$

When the brakes are applied, $a = -4\,\text{m/s}^2$, and the time t to come to rest can be found by using the formula,

$$v = u + at \text{ with } v = 0,\ u = 20\,\text{m/s}$$

$$\text{i.e.} \quad 0 = 20 - 4t, \text{ so } t = 20/4 = 5\,\text{s} \quad \textit{Ans}$$

The braking distance,

$$\begin{aligned} s &= ut + \tfrac{1}{2}at^2 \\ &= (20 \times 5) - (\tfrac{1}{2} \times 4 \times 5^2) \\ &= 100 - 50 \\ &= 50\,\text{m} \quad \textit{Ans} \end{aligned}$$

which, of course, checks with

$$\begin{aligned} s &= \text{average velocity} \times \text{braking time} \\ &= \tfrac{1}{2}(20 + 0) \times 5 = 50\,\text{m} \end{aligned}$$

2 A train travelling at 10 m/s accelerates uniformily over a distance of 5000 m to achieve a velocity of 30 m/s. Calculate its acceleration over this distance and the time taken.

Solution

Applying the formula,

$$v^2 = u^2 + 2as$$

$$\begin{aligned} \text{we have} \quad 30^2 &= 10^2 + (2a \times 5000) \\ 900 - 100 &= 10\,000a \end{aligned}$$

$$\text{so} \quad a = \frac{800}{10\,000} = 0.08\,\text{m/s}^2 \quad \textit{Ans}$$

To find the time t we can use

$$\begin{aligned} v &= u + at \\ \text{i.e.} \quad 30 &= 10 + 0.08t \end{aligned}$$

$$\text{so} \quad t = \frac{30 - 10}{0.08} = 250\,\text{s} \quad \textit{Ans}$$

3 A body with an initial velocity of 5 m/s rolls down an incline with a constant acceleration. If the length of the incline is 120 m and the time taken is 10 s, determine the average velocity, the final velocity and acceleration of the body.

Solution

$$\begin{aligned} \text{Average velocity } \bar{v} &= \frac{\text{length of incline}}{\text{time taken}} \\ &= \frac{120}{10} = 12\,\text{m/s} \quad \textit{Ans} \end{aligned}$$

$$\begin{aligned} \text{But} \quad \bar{v} &= \tfrac{1}{2}(\text{initial} + \text{final velocity}) \\ \text{so} \quad 12 &= \tfrac{1}{2}(5 + v) \\ \text{i.e.} \quad 24 &= 5 + v, \text{ hence } v = 24 - 5 \\ &= 19\,\text{m/s} \quad \textit{Ans} \end{aligned}$$

Using $v = u + at$, we have
$$19 = 5 + (a \times 10)$$
so the acceleration,
$$a = \frac{19 - 5}{10} = 1.4\,\text{m/s}^2 \quad \textit{Ans}$$

6.6 Friction and frictional resistance

Whenever a body is in motion or tending to be forced to be moved there are always forces set up to oppose the motion or oppose the tendency of the body to move. These forces are known as **friction forces.**

Friction forces are resistive forces which always oppose the motion of a body and therefore always act in a direction opposite to that of the motion and the applied force. They oppose the relative motion between two surfaces; for example, when a body is sliding on a surface its motion is impeded by friction forces acting at the surfaces of contact. Friction forces also occur in gases and liquids. For example, when a body falls through air or when a vessel moves through water, frictional resistive forces oppose their motion.

Friction forces between solids are always present at the contacting surfaces of the solids when they slide or tend to slide over each other. The friction forces may be considered as the resistance to motion caused by surface irregularities. Even when the contacting surfaces appear smooth or even highly polished, a closer examination under a microscope will show the surface contours as consisting of a multitude of irregular valleys and peaks. Thus when an applied force acts to set up motion it will be opposed by the resistance produced by these irregularities tending to interlock.

6.6.1 Static and kinetic friction forces for contacting solids

The frictional forces between two contacting solids may be defined on a more quantitative basis with reference to Figure 6.5. A mass of m kilograms rests on a plane surface. The gravitational force on the mass is mg newtons acting vertically downwards. This force is exactly counterbalanced by the normal reaction force N of the plane on the block acting vertically upwards, so $N = mg$. N (or in general the perpendicular force pressing the two surfaces together) and the nature of the surfaces in contact are the primary factors determining the friction force. Next consider a small horizontal force F_a acting on the block but of insufficient magnitude to move the block. Immediately F_a is applied, the friction force F is set up in the plane of contact to resist motion. The friction force $F = F_a$ and this equality remains so as F_a is initially increased, but there comes a point at which the maximum friction force that can exist between the block and the plane is exceeded by F_a. At this point the block will begin to slide.

The friction force which prevents the block from moving is known as the **static friction force**. The maximum value of this force is known as the limiting friction force, which we will denote by F_1. At the point when $F_a = F_1$ sliding is just about to take place. When F_a is greater than F_1 the block will move in the direction of the applied force. Actually, once sliding has been established it is found that the friction force opposing motion is

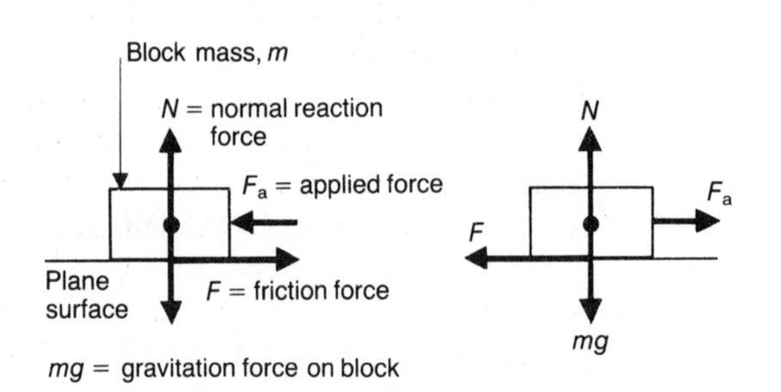

(a) Diagrams showing normal reaction N and friction force F Note that F acts in opposite direction to applied force F_a

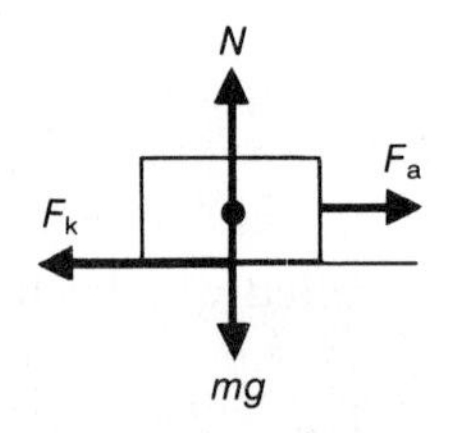

(b) Forces when block in motion Friction force = F_K, the kinetic friction force

Figure 6.5

slightly less than the limiting friction force. The friction force opposing motion when motion has been established is known as the **kinetic friction force**, F_k.

The magnitude of the friction force between two solids in contact and whose areas of contact are clean and dry, i.e. free from water, grease, oil, etc., depends on the following factors:

1 The perpendicular force between the two surfaces of contact, e.g. the normal reaction force N in Figure 6.5.
2 The nature of the materials in contact and their smoothness.
3 The friction force does not normally depend on the actual area of contact, provided that this area is not so small as to cause penetration (e.g. grooving) or seizing which would prevent normal sliding motion.
4 The friction force is normally independent (at least for certain materials and over certain ranges of sliding speeds) of sliding speed, provided that any heat produced by sliding does not change surface conditions.
5 The limiting friction force (maximum force of friction just before sliding occurs) is greater than the kinetic friction force (friction force when motion is established).

For clean, dry surfaces it is possible to form a simple relationship between the kinetic friction force F_k and the perpendicular force N between two given surfaces in uniform sliding motion. The relationship is only approximate and does not apply under all conditions; for example it is inaccurate when N is relatively small and also when N is relatively high. The relationship, based on experimental observations, is

$$F_k = \mu_k N$$

where μ_k (Greek letter mu) is a constant for two given materials, μ_k is known as the coefficient of sliding or kinetic friction:

$$\mu_k = \frac{F_k}{N} = \frac{\text{kinetic friction force}}{\text{perpendicular or normal force between the surfaces}}$$

A similar relationship can also be used to relate limiting friction F_l and N:

Table 6.4 Some average values of coefficients of sliding or kinetic friction for clean, dry surfaces (except for lubricated steel)

Materials	μ_k	*Materials*	μ_k
Wood on wood	0.4	Steel on concrete	0.3
Steel on steel	0.2	Steel on steel (lubricated)	0.03
Steel on ice	0.04	Rubber on concrete	0.75

$$F_l = \mu_s N, \quad \mu_s = \frac{\text{limiting friction, } F_l}{\text{normal force, } N}$$

μ_s is known as the coefficient of static or limiting friction, and since F_l is greater than F_k, μ_s is greater than μ_k. In practice μ_s is anything from a few per cent to as much as 80% higher than μ_k. Some typical values of the coefficient of kinetic friction are given in Table 6.4.

6.6.2 Examples of practical applications and design implications of friction

Everyday examples of the application of friction forces are easy to find: when we strike a match the work done against friction is dissipated as heat and this raises the temperature of the match to ignition point and it bursts into flame; nails, screws, and nuts and bolts rely on friction forces to prevent them from working loose; when we apply the brakes in a car (Figure 6.6) the brake shoes are pressed firmly against the brake drum and the mechanical energy of motion is transformed into heat by the friction forces at the interface between the brake linings and inner drum surface; friction forces are used to transfer mechanical energy to a machine by means of a belt drive or friction clutch (Figure 6.7).

Friction forces between moving surfaces of machines cause wear and the wasteful dissipation of energy as heat. Friction eventually causes either excessive wear or the surfaces become permanently damaged, and the associated heat often causes additional damage to the surfaces as well as presenting the problem of providing adequate cooling to prevent overheating.

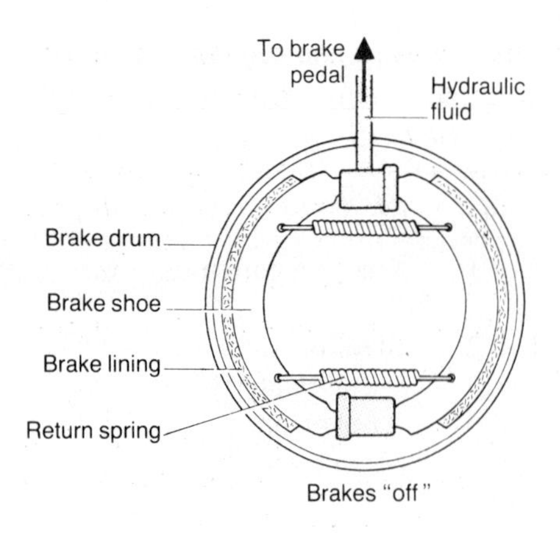

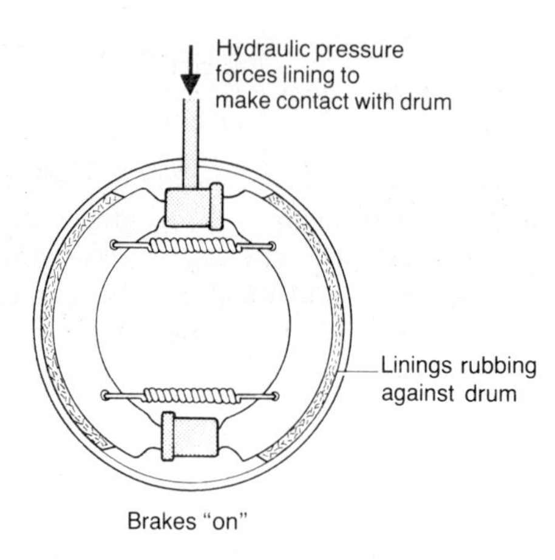

Figure 6.6 Applications of friction: brakes

Thus it is essential in design to reduce friction effects. This will save energy, lengthen machine life, reduce maintenance, and generally lead to improved performance. Friction may be vastly reduced by lubricating contacting metal surfaces by oil or grease, as for example in Table 6.4 which shows that by lubrication the coefficient of kinetic friction between steel surfaces may be reduced from 0.2 to 0.03. In the ideal situation the contacting boundary surface is covered by a thin film of lubricant. This film reduces friction forces compared with 'dry-clean' friction by factors of over 10 to even 100 or more. The friction force for lubricated surfaces is also largely independent of the perpendicular force between the surfaces, but increases with surface area and with sliding speed.

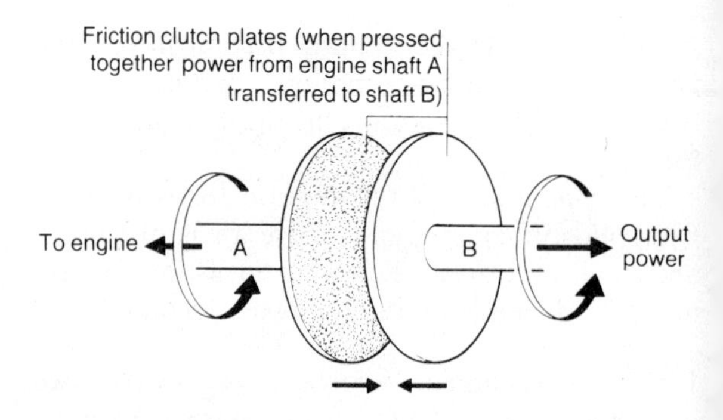

Figure 6.7 Applications of friction: friction clutch

Examples

1 A body of mass 200 kg rests on a horizontal surface. The coefficients of static and kinetic friction between the two contacting surfaces are respectively $\mu_s = 0.25$ and $\mu_k = 0.2$. Calculate the minimum force to move the body and, once the body is moving, the minimum force required to keep the body sliding at a constant speed. Take $g = 9.81\,\text{m/s}^2$.

Solution

The body will just start to move when the applied force is about to exceed the limiting friction force,

$$F_1 = \mu_s N$$

where $N = 200\,\text{g} = 200 \times 9.81 = 1962\,\text{N}$

is the normal reaction force between the 200 kg mass and the surface.

Thus the minimum force required to move the body,

$$F_a = F_1 = 0.25 \times 1962 = 490.5\,\text{N} \quad \textit{Ans}$$

Once the mass is in motion, the minimum force required to maintain a constant sliding speed is equal to the kinetic friction force,

$$F_k = \mu_k N = 0.2 \times 1962 = 392.4\,\text{N} \quad \textit{Ans}$$

2 Figure 6.8 shows a diagram of a steel block of mass 8 kg resting on the surface of a magnetic chuck. The coefficient of static friction between the block and chuck is 0.1. The chuck can be magnetized and the normal reaction

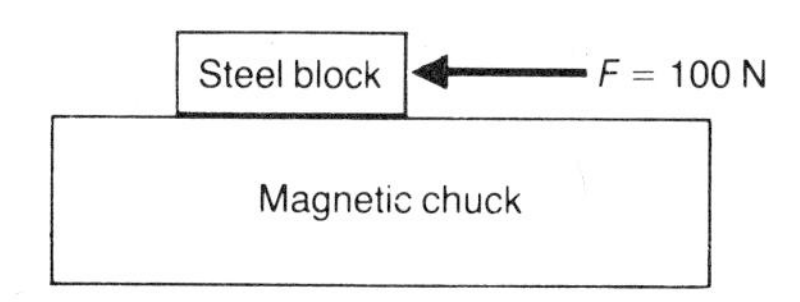

Figure 6.8 For example 2

force N between the block and chuck increased. If the block is subjected to a horizontal force of 100 N, calculate the minimum magnetizing force which must be provided by the chuck to prevent the block moving. Take $g = 10\,\text{m/s}^2$.

Solution

The condition for the block not to move is that the force of limiting friction be greater or equal to the applied 100 N force. To achieve this the minimum magnetization force N_m must satisfy:

$$\mu_s\ (mg + N_m) \geqslant 100\,\text{N}$$

i.e. $0.1\,(8 \times 10 + N_m) = 100$

$$80 + N_m = 1000$$

so $N_m = 920\,\text{N}$ *Ans*

Test 6

This test may be used as a basic self-assessment test to check whether you have absorbed the main facts of Chapter 6 on **Acceleration, velocity–time graphs and solving problems in linear motion**, and its learning objectives. Enter all answers in the answer block.

Qu. 1 Enter a tick (√) in the answer block if you consider the statement is correct; enter a cross (×) if you consider the statment is in any way incorrect.

(a) Acceleration is the rate of change of velocity with time.

(b) The gradient of a velocity–time graph at a given point equals the acceleration at that point.

(c) Gravitational forces act to 'pull' bodies downwards so that if they were free to fall, they would accelerate at $g = 9.81\,\text{m/s}^2$.

(d) Friction forces always act to oppose motion.

(e) A force of 50 N produces an acceleration of $5\,\text{m/s}^2$ on a mass; if the applied force is doubled and the mass is also doubled the acceleration increases to $20\,\text{m/s}^2$.

Qu. 2 Plot the velocity–time graph for the data:

Time (s)	0	1	2	3	4	5	6
Velocity (m/s)	10	10	10	8	6	4	2

and calculate:

(a) the distance travelled in the interval $t = 0$ to 2 s;

(b) the distance travelled in the interval $t = 2$ to 6 s;

(c) the acceleration over the interval $t = 2$ to 6 s;

(d) the average velocity over the total period $t = 0$ to 6 s.

Qu. 3 A stone is dropped from rest from a high tower of height 200 m. Taking $g = 10\,\text{m/s}^2$ and neglecting any resistive forces, calculate

(a) the velocity of the stone after 3 s;

(b) the distance fallen in 3 s;

(c) the time taken for the stone to reach the ground.

Qu. 4 A force of 20 N is applied to a mass of 5 kg for a duration of 10 s. The mass is initially at rest. Calculate:

Answer block:

<table>
<tr><td rowspan="2">Question no.</td><td colspan="5">1</td><td colspan="4">2</td><td colspan="3">3</td><td colspan="3">4</td></tr>
<tr><td>(a)</td><td>(b)</td><td>(c)</td><td>(d)</td><td>(e)</td><td>(a)</td><td>(b)</td><td>(c)</td><td>(d)</td><td>(a)</td><td>(b)</td><td>(c)</td><td>(a)</td><td>(b)</td><td>(c)</td></tr>
<tr><td>Answer</td><td></td><td></td><td></td><td></td><td></td><td></td><td></td><td></td><td></td><td></td><td></td><td></td><td></td><td></td><td></td></tr>
</table>

(a) the acceleration over this period;
(b) the velocity at time $t = 6$ s;
(c) the distance travelled from $t = 6$ to 10 s.

Problems 6

1 (a) Define acceleration and explain how acceleration at a given instant can be determined from a velocity–time graph.
(b) A body moving in a straight line is accelerated at a constant rate of 5 m/s^2 by an applied force of 100 N. What is the mass of the body?
(c) A car braking in a straight line has an initial velocity of 40 m/s and pulls up in 20 s. Assuming the deceleration is constant, determine its value and the total distance travelled by the car in coming to rest.

2 The velocity–time graph of Figure 6.9 describes the motion of a body moving in a straight line. Determine
(a) the body's acceleration in the time intervals:

$t = 0$ to 30 s, $t = 30$ to 80 s,
$t = 80$ to 100 s;

(b) the total distance travelled and the average velocity over the complete time $t = 0$ to 100 s.

3 Plot the velocity–time graph for the data:

Time, t (s)	0	5	10	15	20
Velocity, v (m/s)	100	138	171	192	200

Time, t (s)	25	30	32.5	35	37.5
Velocity, v (m/s)	200	200	121	73	45

Time, t (s)	40	42.5	45	50
Velocity, v (m/s)	27	16	10	0

and estimate:

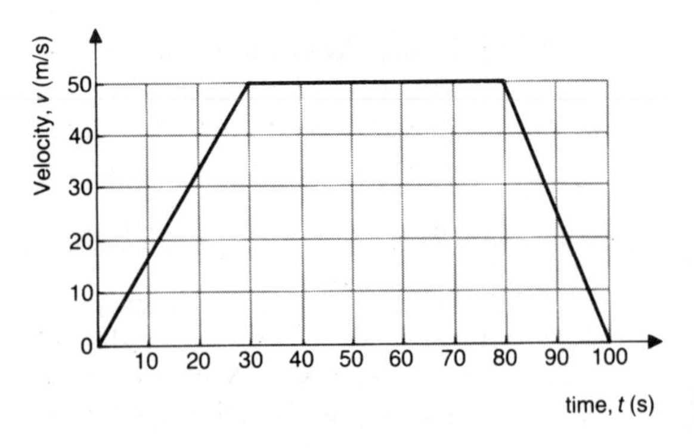

Figure 6.9 v-t *graph for problem 2*

(a) the acceleration at $t = 0$;
(b) the acceleration at $t = 30$;
(c) the total distance travelled and the average velocity over the complete time $t = 0$ to 50 s.

4 A body is dropped from a high building with zero initial velocity and strikes the ground 3.8 s later. Taking $g = 9.81$ m/s^2 and neglecting any retarding forces, calculate:
(a) the velocity at which the body hits the ground;
(b) the height of the building.

5 A stone is projected vertically upwards with a velocity of 30 m/s. Calculate the stone's velocity 1.5 s later and the maximum height reached. Take $g = 10$ m/s^2.

6 A car travelling at an initial velocity of 15 m/s accelerates at a constant rate and reaches a velocity of 35 m/s in 5 s. Calculate the acceleration and the distance travelled in the 5 s.
If the brakes are subsequently applied and it can be assumed they produce a constant deceleration, determine the braking distance if the car is brought to rest in 20 s.

7 A missile is projected vertically upwards into the air and reaches its maximum height in 30 s. Determine the velocity of projection and the maximum height reached. Take $g = 10$ m/s^2.

8 A body of mass 25 kg is lifted vertically by a force of 300 N acting for 5 s. Assuming $g = 10$ m/s^2, calculate:

(a) the resultant force acting on the body during this time;
(b) the distance moved by the body, which may be assumed to be initially at rest, during the 5 s;
(c) the additional upward distance moved if the force is removed after the 5 s.

9 (a) Define with the aid of a diagram the magnitude and direction of the friction force acting when a body is sliding over a surface.
(b) State three factors which influence the magnitude of the friction force.
(c) Give two examples of the practical application of friction.

10 A body of mass 100 kg rests on a horizontal surface. If the coefficient of kinetic friction between the two surfaces is $\mu_k = 0.2$, calculate the force required to slide the body at a constant velocity. Take $g = 10\,m/s^2$.

7 Waves and wave motion

General learning objectives: to describe wave motion and to solve problems involving wave velocity.

7.1 Wave motion and examples of some common waves

The concepts of waves and wave motion are utilized by scientists and engineers to describe and explain and also to predict the behaviour of a wide variety of physical phenomena and processes, all of which have in common the property that energy may be transferred from one point to another in a medium without any net transfer of the medium. At any point along the path of wave motion only a periodic displacement or vibration takes place about a mean position. The energy carried by the wave travels through the medium without actually carrying the medium with it.

For example, when a wave moves on the surface of water, water is not transported bodily with the wave. As the wave propagates, the water molecules actually have a circular or elliptical-type motion, as illustrated in Figure 7.1. When wind produces waves in a field of wheat or corn, the wave travels across the field but the actual corn, of course, remains in place swaying about its mean position.

Some examples of waves and wave motion are listed here.

1 *Surface water waves*
Waves on seas, etc. are perhaps the most visual of all wave motions. The waves generated on the surface of water can propagate vast amounts of energy, which, for example, is expended on our beaches in breaking up rocks and eroding coast lines. Much research is being undertaken currently into practically efficient means of harnessing such energy.

2 *Electromagnetic waves*
Radio waves, heat, light, ultra-violet, X- and gamma-rays are all examples of electromagnetic waves, so called because they consist of combined electric and magnetic field oscillations. Electromagnetic waves in contrast to other waves require no material medium for their transmission and propagate energy through a vacuum. In free space the electric and magnetic fields oscillate at right angles to the direction of propagation (see Figure 7.2) and travel at the velocity of light, $c = 3 \times 10^8$ m/s.

3 *Sound waves*
Any vibrating body in contact with any material medium (solid, liquid, or gas, but not a vacuum) acts as a source of sound waves. The sound waves carry the vibrational energy

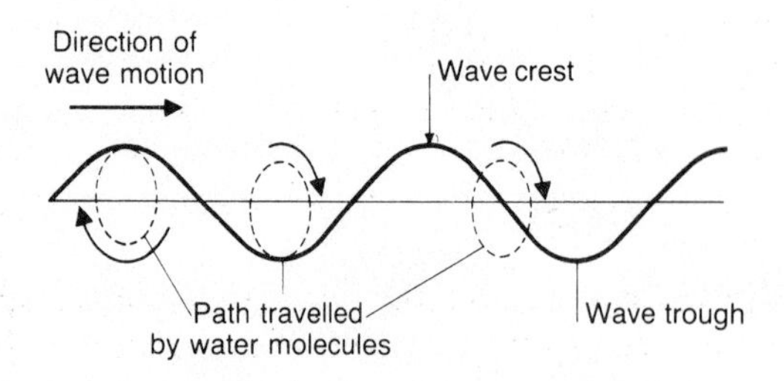

Figure 7.1 Water surface waves

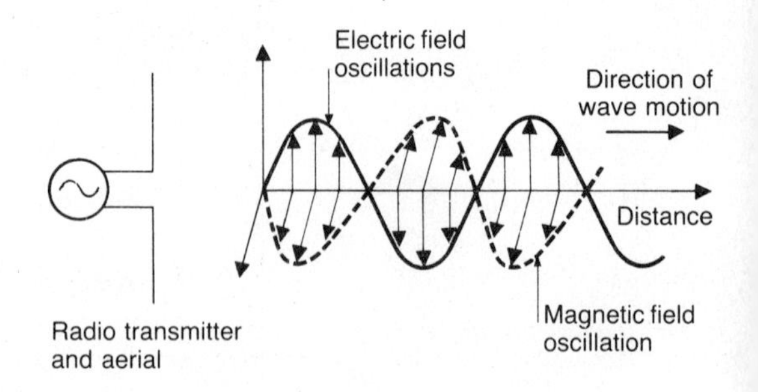

Figure 7.2 Form of electromagnetic wave radiated by a radio transmitter's aerial

imparted to them by the source. They are transmitted through the medium by means of successive condensations (or compressions) and rarefactions of the molecules of the medium. A condensation is caused by the wave packing the molecules more tightly and therefore to a higher pressure than the undisturbed medium. A rarefaction corresponds to the effect of the wave separating the molecules or moving them wider apart and hence creating a lower-pressure region. The soundwave causes a continual oscillation of the molecules about their mean position, and therefore a continual variation of pressure in the medium along the direction the wave travels. We can thus describe a sound wave as a pressure wave which produces variations of pressure in the medium about the average value.

To reinforce the description of sound as a pressure wave let us consider the production and transmission of sound waves in air, the sound in our example being produced by striking a tuning fork. When the fork is set in vibration it compresses and therefore increases the pressure of the air to the outside of the fork prongs as the prongs move in the outward direction and rarefies or reduces the pressure of the air as the prongs move inwards. The maximum pressure corresponds to the instant the prongs are moving outwards with maximum speed, the minimum pressure to the instant when the prongs are moving inwards with maximum speed. These pressure variations are then passed on to neighbouring sections of air, which in turn increase and decrease the pressure of further sections, and so on.

The resulting effect is shown in Figure 7.3 for a given instant of time. The vertical lines represent layers of molecules of air; the closer these lines are together indicates increasing pressure, the more widely the lines are separated indicates reducing pressure below atmospheric. In practice the pressure variations are extremely small, being well below $\pm 0.2\%$ of atmospheric pressure even for the highest energy levels of sound.

Examples of various types of sound wave are: waves generated by speech and musical instruments, which propagate through air, taking with them the energy from the source to our ear; troublesome vibrations transmitted through buildings; seismic waves transmitted through the earth's crust by an earthquake; ultrasonic waves set up in water and used for depth sounding; ultrasonic waves propagated in a material or a component to test for faults within the material or component; noise generated by car and aero engines – the vibrations caused by the latter set up sound waves in air known as noise.

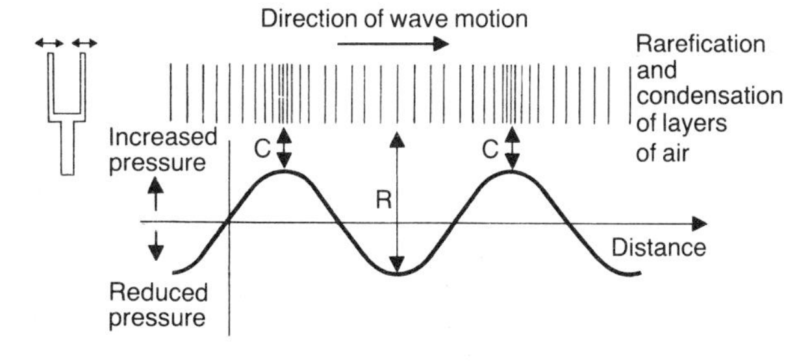

Sound wave generated by a tuning fork
At position C layers of air are mostly 'condensed' (highest pressure)
At position R layers of air are most 'rarified' (lowest pressure)
Graph shows variation of pressure of wave at a given instant of time corresponding to state of layers shown.

Figure 7.3 Nature of sound wave produced by a vibrating tuning fork

7.2 Wavelength

Wavelength is the distance between two successive points in a medium in which a wave is propagating, which are experiencing identical changes, e.g. identical condensation, rarefaction, displacement, etc., and the directions of the change are also identical in the sense that both are either increasing or decreasing. More concisely, we can state that wavelength is the distance between successive points on the wave which are in-phase with each other. Wavelength is denoted by the Greek letter lambda, λ and has the units of metres.

Wave motion and the concept of wavelength can be demonstrated visually in the laboratory by vibrating one end of a loosely coiled spring – a coil of length 3 to 5 m and diameter about 50 mm

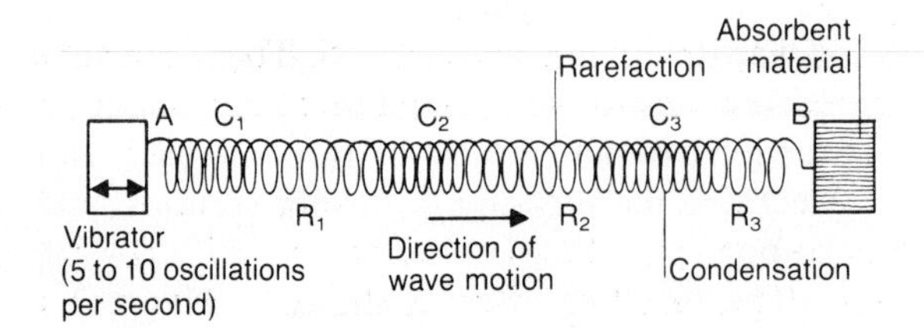

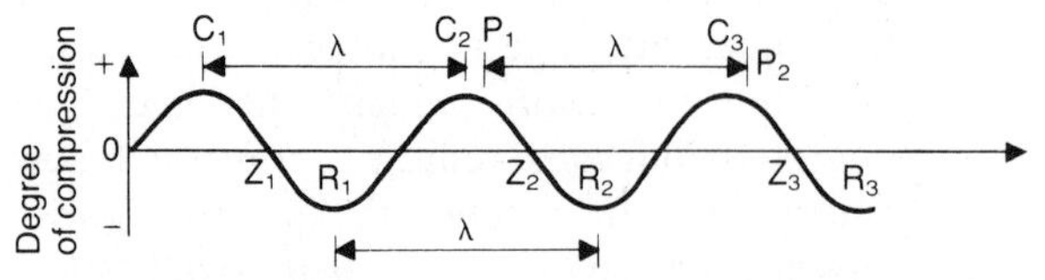

"Photograph" at instant of time spring is in the condition shown above

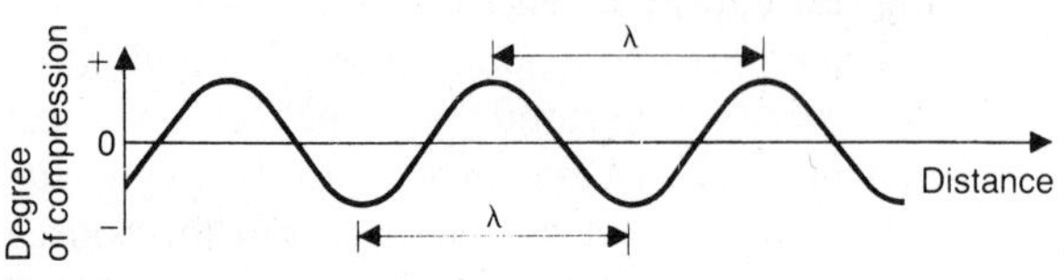

"Photograph" at later instant i.e. wave has moved forward

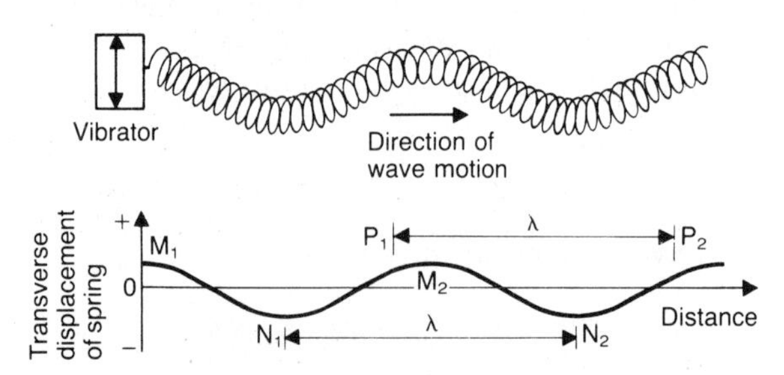

Figure 7.4 Simulation of wave motion and definition of wave length λ
(a) Longitudinal wave motion in a spring (analogous to sound waves)
(b) Simulation of transverse wave motion in a long spring

is suitable. The principle aspects of the demonstration are sketched in Figure 7.4.

The vibrator alternately compresses and extends the spring and these changes are transmitted from the vibrator at end A to travel down the spring where they are absorbed at the far end B by an 'absorbent' load. It can be observed that distances between successive maxima of condensation, C_1 and C_2, or rarification, R_1 and R_2, are always the same, even though they are constantly moving from left to right. If we were to plot the relative packing of spring turns per centimetre versus the distance from end A at a given instant of time, we would obtain a graph of the form shown in (ii). At a slightly later instant of time, we would obtain a graph of exactly the same form but shifted along the distance axis as shown in (iii). We can think of (ii) and (iii) as 'photographs' taken at two relatively close but separate instants of time. We can use these wave diagrams to determine the wavelength λ.

In general λ = distance between successive in-phase points, so referring to the graph of Figure 7.4(a), (ii):

$\lambda = C_1C_2 = C_2C_3$ = distance between two consecutive maxima,

or $\lambda = R_1R_2 = R_2R_3$ = distance between two consecutive minima,

or $\lambda = Z_1Z_2 = Z_2Z_3$ = distance between two consecutive zeros (where the change from + to − (or − to +) is the same),

or $\lambda = P_1P_2$ (two consecutive in-phase points where compressions are identical).

The wave motion simulated in Figure 7.4(a) is known as **longitudinal** wave motion since the spring vibrates in the same direction as the wave travels. In general, longitudinal waves are defined as waves which cause vibration of the medium in the direction in which the waves are travelling. Sound waves are longitudinal waves.

Transverse waves are waves whose displacements are at right angles to the direction at which waves are travelling. Vibrating strings, light waves and, to some extent, surface water waves are examples of transverse waves. Transverse waves are simulated in our long coil by vibrating end A at right angles to the coil axis. This situation is shown in Figure 7.4(b). The graph plotted is of transverse distance moved by the coil from its axis at a given instant of time. The wavelength for this wave is defined identically, i.e.

$\lambda = M_1M_2$ = distance between two consecutive maxima of transverse displacement

$= N_1N_2$ = distance between two consecutive minima of transverse displacement

$= P_1P_2$ (distance between any two consecutive points with identical displacement and displacement change)

Note that the wavelength of the longitudinal and transverse waves given above will not in general be equal.

7.3 Periodic time and frequency

The time taken for one complete cycle of events at any given point in a medium through which a wave is propagating is known as the **periodic time**, often denoted by T.

For example, if we were to focus our attention on a particular section of the spring in the previous spring–vibrator–wave demonstrations and were able to plot the degree of compression or displacement with time, we would obtain graphs similar to those shown in Figure 7.5. These curves show the periodic nature of what would happen. The compression or displacement would undergo a complete cycle from zero through a maximum, decreasing back through zero to a minimum, and so on.

The **frequency** of a wave is the number of cycles of the wave motion occurring in one second. Thus, as one cycle takes T seconds (the periodic time), the number of cycles per second, the frequency

$$f = \frac{1}{T}$$

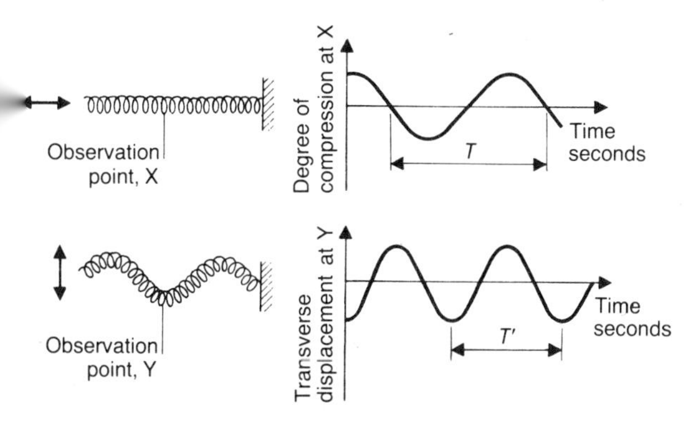

Figure 7.5 Variations taking place at a given point: periodic time = time for one complete cycle of variation

The frequency of a wave is, of course, equal to the frequency of the source creating the wave motion.

The SI unit of frequency is the hertz (Hz), named after Heinrich Hertz (1857–94), the famous scientist who first confirmed the existence of electromagnetic waves. Common multiple units of frequency are:

1 kilohertz, 1kHz = 1000 Hz
1 megahertz, 1MHz = 10^6 Hz
1 gigahertz, 1GHz = 10^9 Hz

7.4 Velocity of waves, $v = f\lambda$

The velocity at which waves travel through a medium is given by the formula,

$$v = f\lambda \text{ metres per second (m/s)}$$

where f = frequency in hertz (the number of cycles per second)
λ = wavelength in metres

This result can easily be established. The wave motion moves forward a distance equal to the wavelength in the time taken for the vibration or displacement at any given point to have undergone one complete cycle about its mean position. Thus a distance λ is moved in the periodic time of T, so

$$\text{wave velocity, } v = \frac{\text{distance travelled}}{\text{time taken}} = \frac{\lambda}{T} = \lambda \times \left(\frac{1}{T}\right) = \lambda f$$

as $T = 1/f$ or $1/T = f$

The velocity of electromagnetic waves in a vacuum (in free space) is 2.9979×10^8 m/s and is normally approximated to 3×10^8 m/s with little error. Thus radio waves and light travel pretty rapidly in space to say the least! Electromagnetic waves travel somewhat slower in dielectric media. For example, in glass the velocity is reduced by a factor of about 1.5 (the refractive index), in polythene (the material frequently used as the insulator in coaxial lines) by the same factor of 1.5, while in water the waves are slowed by a factor of 1.33.

The velocity of sound depends on the medium or material, or more specifically on the density

Table 7.1 Velocity of sound in some common media

Medium	*Velocity (m/s)* (T in °C)
Air	$331.46 + 0.61T$
Water	$1403 + 4.2T - 0.028T^2$
Sea-water	$1449 + 3.6T$
Aluminium	5100
Copper	3600
Steel	5100
Wood	3000–4000
Glass	5000–6000
Brick	≈3700

and the elastic properties of the medium or material. For example the velocity of sound in air (at 0°C) is 331.5 m/s, the velocity of sound in steel is 5100 m/s, which explains why we can 'hear' an approaching train first through the railway lines before we 'hear' it through the air. The velocity of sound depends on the medium and varies to some extent with temperature. It does not depend on the frequency or magnitude of the sound energy being propagated. Values of the velocity of sound through air, water, and some solid substances are given in Table 7.1.

7.5 The solution of problems involving wave velocity

Remember, wave velocity $v = f\lambda$ m/s
where f = frequency (number of cycles per second), hertz (Hz)
λ = wavelength, metres (m)

so $f = \dfrac{v}{\lambda}$ and $\lambda = \dfrac{v}{f}$

Examples

1 In an experiment to measure the velocity of electromagnetic waves in air the following results were obtained:

wavelength $\lambda = 16.17$ mm
frequency $f = 18.54$ GHz

Calculate the velocity of electromagnetic waves in air to four significant figures.

Solution

$\lambda = 16.17\,\text{mm} = 16.17 \times 10^{-3}\,\text{m}$
$f = 18.54\,\text{GHz} = 18.54 \times 10^{9}\,\text{Hz}$
so velocity, $v = f\lambda$
$= 18.54 \times 10^{9} \times 16.17 \times 10^{-3}$
$= 299.79 \times 10^{6}$
$= 2.998 \times 10^{8}\,\text{m/s}$ *Ans*

2 A diaphragm vibrating at a frequency of 256 Hz produces a sound wave in air of wavelength of 1.294 m. Calculate the velocity of sound in air.

Solution

$\lambda = 1.294\,\text{m}, f = 256\,\text{Hz}$
so velocity $v = f\lambda = 256 \times 1.294$
$= 331.3\,\text{m/s}$ *Ans*

3 Figure 7.6 shows the side view of water waves travelling in a tank. The waves are generated by a vibrating metal strip.
(a) Determine the wavelength of the waves.
(b) If crest A takes 2 s to travel to B, calculate the wave velocity and frequency of the vibrating metal strip.

Solution

Distance AB $= 2\frac{1}{4}\lambda$
since distance $AX_1 = \lambda$, $X_1X_2 = \lambda$ and $X_2B = \frac{1}{4}\lambda$
hence $2.25\lambda = 0.81\,\text{m}$
therefore $\lambda = \dfrac{0.81}{2.25} = 0.36\,\text{m}$ *Ans*

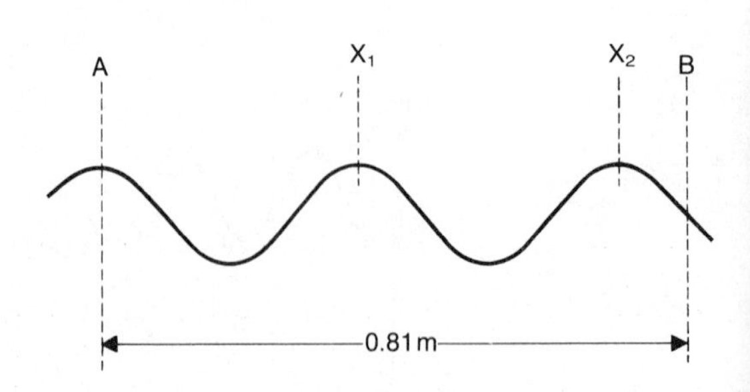

Figure 7.6 Wave diagram for example 3

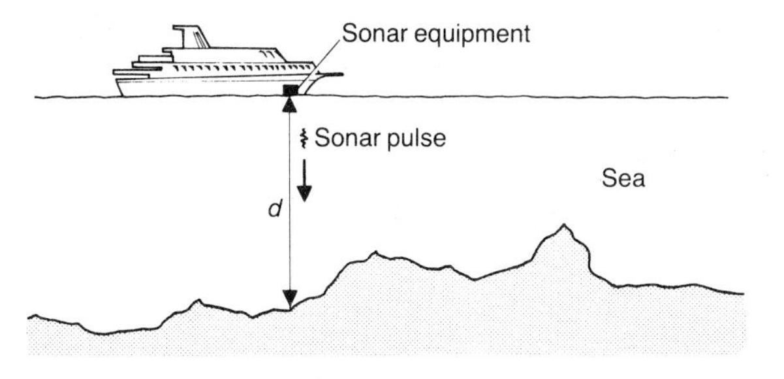

Figure 7.7 Sonar depth sounding (see example 4)

Since crest A travels 0.81 m in 2 s, the wave velocity

$$v = \frac{\text{distance}}{\text{time}} = \frac{0.81}{2} = 0.405\,\text{m/s} \quad \textit{Ans}$$

The frequency of the wave and hence the frequency of vibration of the metal strip producing the wave is

$$f = \frac{v}{\lambda} = \frac{0.405}{0.36} = 1.125\,\text{Hz} \quad \textit{Ans}$$

4 A ship employing sonar equipment (sonar stands for sound navigation and ranging) sends out a pulse of ultrasonic waves at a frequency of 50 kHz, as shown in Figure 7.7. The pulse is reflected from the sea-bed and is picked up by the sonar; the total transit time, from ship to sea-bed and back again is measured. If the velocity of the ultrasonic waves in sea water is 1450 m/s and the transit time is 0.22 s, determine the depth of sea below the ship.

Solution

Let depth of sea below ship be d metres, then the total distance travelled by the ultrasonic pulse of waves is $2d$. Hence

as distance = velocity × time

$$2d = (1450\,\text{m/s}) \times (0.22\,\text{s}) = 319\,\text{m}$$

$$d = 159.5\,\text{m} \quad \textit{Ans}$$

Test 7

This test may be used as a basic self-assessment test to check whether you have absorbed the main facts of Chapter 7 on **Waves and wave motion**, and its learning objectives. All answers to be entered in the answer block.

Qu. 1 Enter a tick (√) in the answer block if you consider the statement is correct; enter a cross (×) if you consider the statement is in any way incorrect.

(a) The wavelength is the distance between successive in-phase points, e.g. between successive maxima.

(b) The frequency of a wave is the number of cycles per second and has the SI unit of hertz.

(c) Wave velocity $v = f\lambda$, where f = frequency, λ = wavelength.

(d) Sound waves can propagate through a vacuum and do so at a faster rate than in air.

Qu. 2 Given that the velocity of electromagnetic waves is 3×10^8 m/s, calculate:

(a) the wavelength of a wave of frequency 1 MHz;

(b) the frequency of a microwave signal from a satellite which has a wavelength of 25 mm.

Qu. 3 (a) Given that sound waves travel at a velocity of 330 m/s, calculate the time for a gun-shot to be heard after firing from a distance 1 km away.

(b) It is observed that when a steel rail is

Answer block:

Question no.	1				2		3	
	(a)	(b)	(c)	(d)	(a)	(b)	(a)	(b)
Answer								

vibrated at a frequency of 2 kHz the time for the transmission of the vibration through a 500 m length of the rail is 0.1 s. Calculate the velocity of the sound vibrations in steel.

Problems 7

1 (a) Explain what is meant by wave motion and give four practical examples.
(b) Explain with aid of diagrams the terms wavelength and frequency.
(c) Draw graphs showing:
(i) the variation of pressure amplitude in a sound wave with distance from the source at a given instant of time;
(ii) the variation of pressure amplitude in a sound wave with time at a given distance from the source.
Mark on the relevant graph the wavelength and the periodic time of the wave.

2 State the relationship between velocity, frequency and wavelength for waves.
A tuning fork vibrating at a frequency of 512 Hz produces a sound wave of wavelength 0.647 m. Calculate the velocity of sound in air.

3 Table 7.2 gives the classification of the frequency ranges for the various radio communication bands. Given that electromagnetic waves propagate in free space at the velocity, $c = 3 \times 10^8$ m/s, determine the corresponding wavelength ranges for each band.

Table 7.2 Classification of radio bands

Frequency range	*Classification*
30 kHz–300 kHz	Low frequency (lf)
300 kHz–3 MHz	Medium frequency (mf)
3 MHz–30 MHz	High frequency (hf) or short wave
30 MHz–300 MHz	Very high frequency (vhf)
300 MHz–3 GHz	Ultra high frequency (uhf)
1 GHz–300 GHz	Microwaves

4 An ultrasonic generator operating at 50 kHz sends waves through sea water at a velocity of 1450 m/s. Determine the wavelength of these waves and time to travel

(i) 10 m; (ii) 500 m.

8 Work, energy and power

General learning objectives: to understand the concepts of work, energy and power and to solve problems associated with these quantities.

8.1 Work and energy: an introduction

In scientific terms we define work in terms of an applied force multiplied by the distance through which the force moves. Thus for work to be performed a force must actually move through a distance. Work, as we shall see, is best represented as the area beneath a force–distance graph.

Some examples of doing work are: lifting a weight from a lower to a higher position – work is done against the force of gravitational attraction; when a material is extended or compressed work is done to overcome the elastic or cohesive forces; a moving car or train is doing work to overcome friction forces and if they move up an incline work is done to overcome both friction and gravitational forces. However, if we were to push against an immoveable object, such as a wall, no work is done in scientific terms – even though we may feel that we are expending a considerable amount of energy.

To provide the force and to do work by moving the force through a distance requires a source of energy. Energy is the capacity to do work.

When we lift a weight energy is expended which enables our muscles to move and do the work. The actual energy supplied to our muscles comes in directly from the food we eat. Most work in an industrial society is accomplished by machines, whose energy is supplied in mechanical, electrical or other forms, but whose source of input is normally derived via heat energy supplied by solid, liquid or nuclear fuel.

Fuels, such as coal, oil, natural gas, and wood are sources of energy. In fact, they are sources which store chemical energy. When they are burnt in air, they liberate heat – heat is also a form of energy – and the heat energy may be applied to do useful work. For example, coal is burnt in power stations to heat water to generate steam. The steam is used to drive turbines to generate electricity – yet another form of energy – and the electricity is distributed to industry and our homes to provide energy to do useful work.

A list of heat energy values of some common fuels is given in Table 8.1. This table gives some guide to the relative energy contained in various fuels which is given out as heat when the fuel is completely burnt.

Table 8.1 Heat energy in joules per kilogram of some common fuels

Fuel	*Heat energy given out in burning 1 kg*
Coal	33×10^6 J*
Oil	45×10^6 J
Natural gas (methane)	56×10^6 J
Wood	15×10^6 J
Petrol	47×10^6 J

Joules (J), the SI unit of energy

8.2 Definition of work in terms of force applied and distance moved

The **work done** by a constant force when it moves its point of application is equal to the product of the force and the distance moved in the direction of the line of action of the force, i.e.

$$W = F \times s$$

where W = work done,
F = force in newtons (N),
s = distance moved in metres (m)

The SI derived unit of work and energy is the

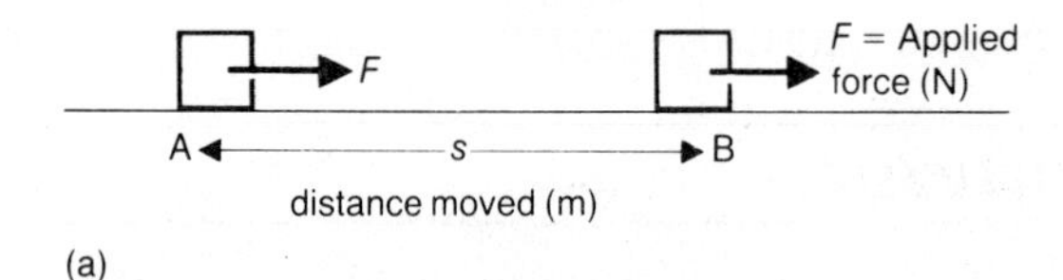

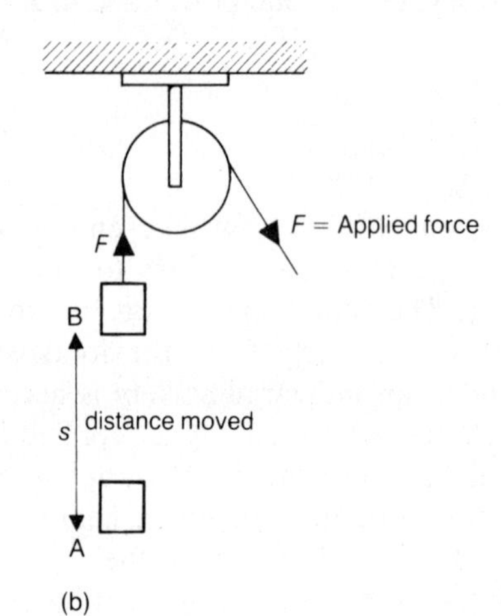

Figure 8.1 Work done = applied force × distance moved

joule (J). The joule is the amount of work which is done when the point of application of a force of 1 N is moved through a distance of 1 m in the direction of the force; 1 J = 1 Nm.

In Figure 8.1(a) and (b) the work done by the applied force F newtons in moving the body from position A to B, a distance of s metres is, for both cases

$$W = F \times s \text{ joules}$$

In case (b), $F = mg$, the gravitational force and if we neglect friction forces which may occur in the pulley system we have

$$W = mg \times s = mgs$$

where m = mass of load in kilograms
$g = 9.81\ \text{m/s}^2$, value of acceleration due to gravity

If the direction of the applied force F and the direction in which the body moves are not in line (see Figure 8.2), then the work done is

$$W = F \times s \cos\theta \text{ joules}$$

where θ = angle between the line of action of the force and the direction the body is moved.

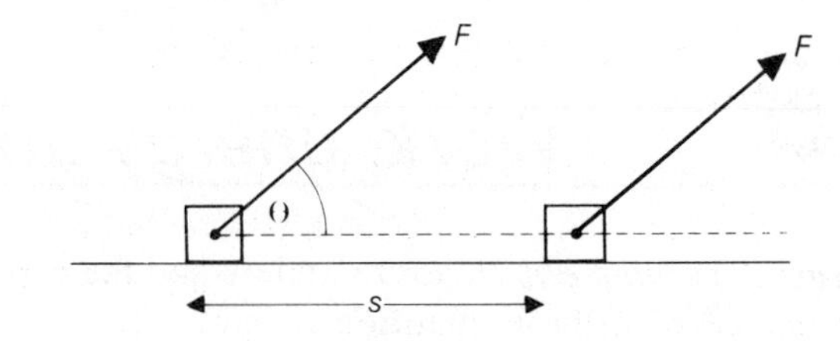

Figure 8.2 Work done, W = F s cos θ

(Trigonometry reference: *Mathematics: A First Course*, by the same author, also published by Hutchinson Education)

Examples

1 Calculate the work done when a force of 25 N moves a body (and therefore its point of application) through a distance of 30 m in the direction of the line of action of the force.

Solution

Work done = force × distance
$= 25 \times 30 = 750\ \text{J}$ *Ans*

2 Calculate the work done in lifting a mass of 500 kg through a vertical height of 28 m. Take $g = 9.81\ \text{m/s}^2$.

Solution

The force F required to lift mass of 500 kg is

$$F = mg = 500 \times 9.81 = 4905\ \text{N}$$

and so the work done in lifting the mass through 28 m is

$$W = 4905 \times 28 = 1.373 \times 10^5\ \text{J or } 0.137\ \text{MJ}$$ *Ans*

3 A sledge is pulled along horizontal ground by a rope which makes an angle of 40° with the horizontal. If the tension in the rope, i.e. the applied force, is 80 N, calculate the work done in moving the sledge a distance of 1500 m.

Solution

Work done, $W = Fs\cos\theta$
where $F = 80\ \text{N}$, $s = 1500\ \text{m}$ and $\theta = 40°$, so
$W = 80 \times 1500 \times \cos 40°$
$= 80 \times 1500 \times 0.7660 = 91920$
or 91.920 kJ *Ans*

8.3 Force–distance graphs and calculation of work done

When dealing with bodies in motion we plotted velocity–time graphs and by calculating the area under the curve we could find the total distance travelled and deduce also the average velocity.

In a similar manner we can plot force–distance graphs and determine the total work done and the average force acting:

Area under force–distance graph = total work done
and if $\bar{F}$ = average force, we have

$$W = \bar{F} \times s$$

$$\bar{F} = \frac{W}{s} = \frac{\text{total work done}}{\text{distance moved}}$$

Examples

1 Plot the force–distance graph for the case shown in Figure 8.3 where a body is moved in three stages:

A to B: distance $s_1 = 30\,\text{m}$; applied force $F_1 = 100\,\text{N}$
B to C: distance $s_2 = 40\,\text{m}$; applied force $F_2 = 70\,\text{N}$
C to D: distance $s_3 = 50\,\text{m}$; applied force $F_3 = 35\,\text{N}$

and determine the work done for each stage, the total work done and the average force acting over the total distance of 120 m.

Solution

The force–distance curve for the complete three-stage movement is drawn in Figure 8.4.

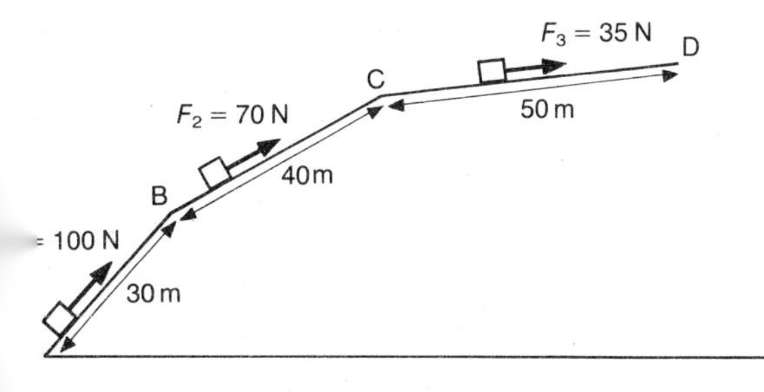

igure 8.3 Forces acting on a body to hold it up three clines AB, BC, and CD (See example 1)

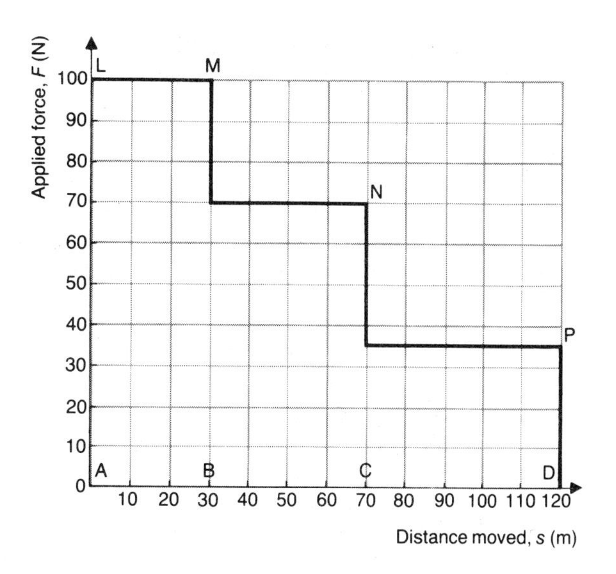

Figure 8.4 Force–distance graph for example 1

From A to B, work done:

$$W_1 = F_1 \times s_1 = 100 \times 30 = 3000\,\text{J} \quad \textit{Ans}$$

Note: W_1 = area LMBA = area under force–distance graph over range 0 to 30 m.

From B to C, work done:

$$W_2 = F_2 \times s_2 = 70 \times 40 = 2800\,\text{J} \quad \textit{Ans}$$
= Area MNCB

From C to D, work done:

$$W_3 = F_3 \times s_3 = 35 \times 50 = 1750\,\text{J} \quad \textit{Ans}$$
= Area NPDC

The total work done,

$$W = W_1 + W_2 + W_3 = 3000 + 2800 + 1750 = 7550\,\text{J} \quad \textit{Ans}$$
= Area under F–s curve

Average force $\bar{F}$ acting over the 120 m, A to D, is

$$\bar{F} = \frac{W}{s} = \frac{7550}{120} = 62.92\,\text{N} \quad \textit{Ans}$$

2 The following experimental results were obtained in an extension test on a metal rod.

Force, F (N)	0	200	400
Extension, x (mm)	0	2.0	4.1

Force, F (N)	600	800	1000
Extension, x (mm)	5.98	8.05	10.0

Plot the force–extension graph of hence determine the work done in extending the rod by 10 mm.

Solution

The force–distance graph is plotted in Figure 8.5, distance in this case being the extension caused by the applied force. The total work done,

$$\begin{aligned} W &= \text{area under force–distance graph} \\ &= \text{area } \triangle \text{ ABC} \\ &= \tfrac{1}{2}\,\text{BC} \times \text{AC} = \tfrac{1}{2} \times 1000\,\text{N} \times 10\,\text{mm} \\ &= 5000\,\text{Nmm} = 5000 \times 10^{-3}\,\text{Nm} \\ &= 5\,\text{Nm or } 5\,\text{J} \quad \textit{Ans} \end{aligned}$$

Note: the extension is in mm so we multiply by 10^{-3} to convert to metres, 1 Nm = 1 J

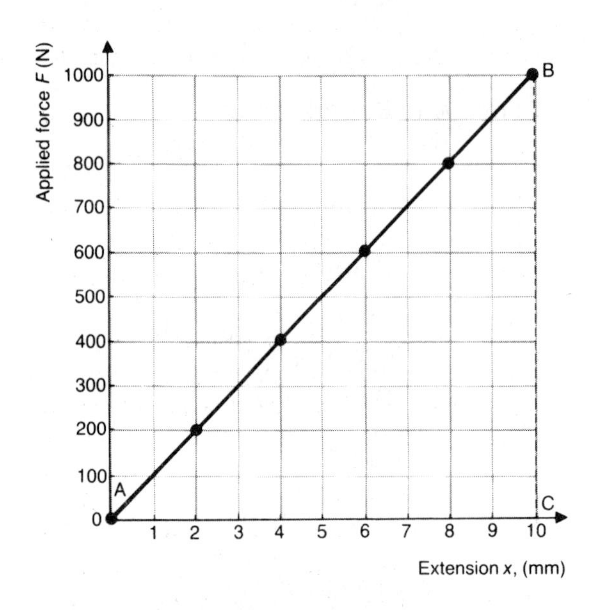

Figure 8.5 Force–distance graph for rod undergoing tensile stress by applied force

8.4 Forms of energy

The energy supplied for work to be done can be regarded as consisting of two basic classes: stored energy, known as **potential energy** and energy which bodies possess by virtue of their motion, known as **kinetic energy**.

Potential and kinetic energy exist in several different forms depending on the source or origin from which the energy is drawn. The main forms of energy are considered briefly below.

1 *Chemical energy*
This is a form of potential energy stored in substances such as coal or petroleum fuels. When the fuel is burnt, the stored chemical energy is converted to heat energy. Chemical energy is also stored in primary and secondary cells (see Chapter 13), which act as sources of electrical energy.

2 *Nuclear (atomic) energy*
Nuclear energy is potential energy stored within the nuclei of atoms. In a nuclear reactor, a tiny quantity of matter is converted into energy by the fission or fusion of atomic nuclei. In the fission process, uranium nuclei are bombarded by tiny uncharged particles known as neutrons and energy is released because the nuclei so formed have less mass than the original uranium nuclei. The difference in mass m is converted to an amount of energy E according to Einstein's famous equation:

$$E = mc^2 \text{ joules}$$

where c = velocity of light = 3×10^8 m/s, m is in kilograms.

The energy released is mainly kinetic energy resulting from the tremendous speeds at which the nuclei formed in the fission process fly apart. If uncontrolled, this kinetic energy provides the destructive force obtained in a nuclear explosion; when controlled, this energy is used to produce heat energy to generate steam. Subsequently the steam may be used to drive a turbine, which in turn drives a generator to produce electrical energy or to provide mechanical energy to drive, for example, a nuclear submarine.

In the fusion process, two light nuclei, such as isotopes of hydrogen known as deuterium combine to form helium, with a resulting loss in mass and hence a release of energy. Fusion processes account for the production of energy in our sun and the stars of the universe.

3 *Heat or thermal energy*
Heat is a form of kinetic energy and

released in most chemical and all nuclear reactions. It is also generated when any work is done against frictional or resistive forces. It is exhibited as an energy source when heat is used to produce steam to provide sources of mechanical or electrical energy, or when petrol or diesel oil is ignited in a combustion engine.

4 *Mechanical energy*
Mechanical energy is a form of kinetic energy. The energy contained in rotating flywheels and turbines provides examples of sources of mechanical energy. In an electrical generator, a coil is rotated in a magnetic field to produce a source of electrical energy; or vice versa in the electrical motor when we pass current through the coil (which requires electrical energy to be supplied) and produce mechanical energy. Examples of the latter are electrical drilling machines and lathes.

Electrical energy
In an electrical generating station, heat is converted to mechanical energy and this mechanical energy is used to produce electrical energy. A vast network of cables and lines (the grid system) transmits the electrical energy from the stations to industry and our homes. Electrical energy may also be obtained from batteries and accumulators by the conversion of the stored chemical energy in the latter. Electrical energy is perhaps the most practical and transportable of energies. We utilize electrical energy to drive machines, to provide heating and lighting, and to provide the power for our radios, televisions, etc.

Light and radiation energy
The principal source of energy on our planet comes from electromagnetic radiation energy produced by the sun. The sun, in fact, generates this energy by nuclear fusion. Electromagnetic energy is radiated through space and it has been calculated that on average approximately 14 000 J of energy are received on the earth per square metre in each second.

The energy contained in the sun's radiation is a key factor in the growth of plant life, and hence in food production. Light energy was also directly responsible, although millions of years ago, for the production of the fuels of coal, oil, and gas (when plant and animal life decayed, they eventually formed these fuels).

Heat, light, and radio waves are forms of electromagnetic radiation. Light transmits energy, enabling us to see. Radio waves provide the means for the transmission of energy from point to point or over large areas, enabling us to use the telephone, listen to the radio, and watch television. (Over longer distances than immediately local calls, telephone messages are carried by radio waves transmitted via satellites or in coaxial cables buried in the ground or under the sea.)

8.5 Energy conversion

Energy can be converted from one form to another as illustrated in the following examples.

1 *Conversion of potential energy (nuclear or chemical) to heat energy to mechanical energy to electrical energy*
The block diagram of Figure 8.6 shows the energy conversion processes involved.

Initially, heat energy is released from the chemical energy stored in a fuel. The heat energy is then used to produce steam or heat gases (i.e. to increase the kinetic energy of water and/or gas molecules) to enable them to do work and drive, for example, a steam or gas turbine, or a piston in the case of an internal combustion engine. Thus, heat energy is converted to mechanical energy. The mechanical energy stored, for example, in a revolving turbine, may then be further converted in an electromechanical generator to produce electrical energy.

2 *Examples of the conversion of mechanical energy to heat energy*
All moving bodies and moving parts of machines experience frictional or similar resistive forces which restrict their motion. Thus, part of the mechanical energy of motion (i.e. kinetic energy) is always used to do work against these forces and is converted to heat. For example, the mechanical energy of a car in motion is reduced and converted to heat

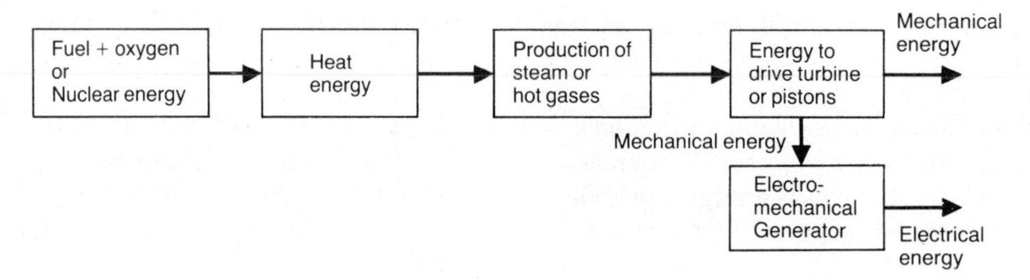

Figure 8.6 A typical energy conversion process for the production of mechanical and electrical energy

energy when the brakes are applied. The brake shoes or discs press against metal surfaces and work is done against the frictional forces at the interface of the brake and metal materials, heat being generated, and thus the mechanical energy contained in the motion of the car is reduced.

Heat is also generated by friction when a meteorite enters the earth's atmosphere. The kinetic energy of the meteorite is reduced at the expense of generating heat. The meteorite not only slows down very rapidly, but also burns itself away. Thus, it is extremely rare to find on the earth's surface a large meteorite. The development of special materials able to withstand the immense heat generation and associated high temperature involved when space vehicles re-enter the earth's atmosphere forms an important part of space research programmes in Russia and America.

3 *Examples of the conversion of electrical energy to heat*
When electrical current flows in a conductor, work is done against the resistance presented by the conductor and electrical energy is converted to heat. Practical examples of the conversion of electrical energy to heat are the electric fire and the electric water heater.

8.5.1 Principles of conservation of energy and mass

When energy is converted from one form to another, no energy is actually lost nor is any extra energy created. Thus in all conversion processes (other than nuclear reactions) we can state: energy cannot be created nor can it be destroyed. This state is known as the *principle of conservation of energy*.

This principle does not infer, for example, that all input energy is converted to a useful work output. Whenever energy in one form apparently disappears, other forms are generated by conversion in exactly equal amounts to that apparently 'lost'.

With the advent of nuclear power, we now know there is a relationship between mass and energy and that in a nuclear reaction, mass can be converted to energy. Prior to this knowledge it was thought that mass as well as energy was conserved. The laws of conservation have now been combined into the *principle of conservation of mass and energy*, which states:

> The total energy and mass in any closed system is conserved (i.e. remains constant and cannot change).

8.6 Efficiency: the ratio of output to input energy (or power)

A practical machine may be regarded essentially as an energy converter. A machine can be defined as any arrangement whose purpose is to take some definite form of energy, modify it and deliver energy in a form more suitable for the desired requirement. Energy of one kind is supplied at the input and energy of other kinds or the same kind is generated and supplied at the output.

In the process of energy conversion, some energy is inevitably wasted and not put to useful work. Machines have certain components, such as bearings, gears and other moving parts, which have to overcome frictional or similar resistive forces in performing their function. Work done against these forces causes part of the input energy supplied to the machine to be dissipated as heat. Consequently, the energy output from a machine is always less than the energy input, i.e.

$$E_{out} = E_{in} - E_{loss}$$

The efficiency of a machine or process, normally denoted by the Greek letter eta η, is defined as the ratio of the output energy to the input energy:

$$\eta = \frac{\text{Output energy}}{\text{Input energy}} = \frac{E_{out}}{E_{in}} = \frac{E_{in} - E_{loss}}{E_{in}}$$

where E_{out} = output energy delivered by machine
E_{in} = input energy supplied to the machine
E_{loss} = energy dissipated in machine

Since energy is always to some extent dissipated in a machine or process and therefore lost, the efficiency is always less than unity.

Examples

1 Determine the efficiency of a machine which provides an energy output of 3000 J when supplied with an energy input of 4000 J.

Solution

Energy output, E_{out} = 3000 J; energy input, E_{in} = 4000 J so

$$\text{Efficiency, } \eta = \frac{E_{out}}{E_{in}} = \frac{3000}{4000} = 0.75 \quad \textit{Ans}$$

Note, efficiency is often quoted as a percentage, i.e.

$$\text{Efficiency} = \frac{E_{out}}{E_{in}} \times 100 \text{ per cent}$$

so in the case above,

$$\eta = 0.75 \times 100 = 75 \text{ per cent}$$

2 Determine the energy input to a machine of efficiency sixty per cent (i.e. $\eta = 0.6$) to provide an output energy of 9000 J.

Solution

$$\eta = 0.6 = \frac{E_{out}}{E_{in}} = \frac{9000}{E_{in}}$$

$$\text{so } 0.6\,E_{in} = 9000$$

$$E_{in} = \frac{9000}{0.6} = 15\,000\text{ J} \quad \textit{Ans}$$

8.7 Power

Power is defined as the rate at which energy is transferred. Thus, if

E = an amount of energy, joules (J)
t = time in seconds in which this energy is transferred, i.e. input to or output from a machine

$$\text{then Power, } P = \frac{E}{t}$$

If the rate of transfer of energy with time is constant then the power P is also constant; if the rate varies then $P = E/t$ is the average or mean power.

The derived SI unit of power, equal to one joule per second, is the *watt* (W):

1 W = 1 J/s

Useful multiple and sub-multiple units of power are:

1 Gigawatt (GW) = 10^9 W
1 Megawatt (MW) = 10^6 W
1 kilowatt (kW) = 10^3 W
1 milliwatt (mW) = 10^{-3} W
1 microwatt (μW) = 10^{-6} W
1 nanowatt (nW) = 10^{-9} W
1 picowatt (pW) = 10^{-12} W

In practice we are concerned primarily with the rate of supply and use of energy, i.e. power, rather than total energies, since power is the quantity which dictates the time required to accomplish the actual tasks. Efficiency can also be defined in terms of power:

$$\text{Efficiency } \eta = \frac{P_{out}}{P_{in}}$$

where P_{out} = output power delivered, W
P_{in} = input power supplied, W

Examples

1 The power input to a machine of 80% efficiency is 1 kW (1000 W). Calculate the output power and the total energy transferred in 30 minutes
(a) to do useful work;
(b) to overcome losses in the machine.

Solution

(a) The output power from the machine,

$$P_{out} = \eta P_{in} = 0.8 \times 1000 = 800\,\text{W}$$

and as *energy = power × time*, the energy transferred for useful work

$$W_{out} = P_{out} \times t$$

where $P_{out} = 800\,\text{W}$

$t = 30 \text{ minutes} = 30 \times 60\,\text{s}$

so $W_{out} = 800 \times 30 \times 60 = 1.44 \times 10^6\,\text{J}$ or $1.44\,\text{MJ}$ *Ans*

(b) The total energy input to the machine

$$W_{in} = P_{in} \times t = 1000 \times 30 \times 60 = 1.8 \times 10^6\,\text{J} \text{ or } 1.8\,\text{MJ}$$

so the total energy used to overcome losses in the 30 minutes is

$$W_{loss} = W_{in} - W_{out} = 1.8 - 1.44 = 0.36\,\text{MJ}$$ *Ans*

2 Calculate the work done and the output power of an electric motor required to lift a mass of 200 kg through a height of 50 m in 60 s. Take $g = 9.81\,\text{m/s}^2$.
If the power input to the motor is 2.5 kW, calculate its efficiency in the lifting process.

Solution

The force required to lift the 200 kg against gravity is

$$F = mg = 200 \times 9.81 = 1962\,\text{N}$$

and the total work done in lifting the mass through 50 m,

$$W = F \times \text{distance} = 1962 \times 50 = 98\,100\,\text{J} \text{ or } 98.1\,\text{kJ}$$ *Ans*

The power required,

$$P_{out} = \frac{W}{t} = \frac{98\,100}{60} = 1635\,\text{W} \text{ or } 1.635\,\text{kW}$$ *Ans*

The efficiency of the motor,

$$\eta = \frac{P_{out}}{P_{in}} = \frac{1.635}{2.5} = 0.654 \text{ or } 65.4\%$$ *Ans*

Test 8

This test may be used as a basic self-assessment test to check whether you have absorbed the main facts of Chapter 8 on **Work, energy and power**, and its learning objectives. All answers to be entered in the answer block.

Qu. 1 Enter a tick (√) in the answer block if you consider the statement is correct; enter a cross (×) if you consider the statement is in any way incorrect.
(a) Work is defined as equal to force applied times distance moved by the force in the direction of its line of action.
(b) Power = work × time and has the unit of watts.
(c) The efficiency of a machine is defined as the ratio of energy output to energy input.
(d) Energy can be converted from one form to another, e.g. electrical energy can be converted to mechanical energy.

Answer block:

Question no.	1					2			3		4		
	(a)	(b)	(c)	(d)	(e)	(a)	(b)	(c)	(a)	(b)	(a)	(b)	(c)
Answer													

(e) Potential energy is stored energy, whilst kinetic energy is energy due to a body's motion.

Qu. 2 Calculate the work done when:
(a) a force of 40 N moves a body a distance of 5 m in the direction of the force;
(b) a weight of 50 N is lifted through a height of 7 m;
(c) a force of magnitude 100 N and making an angle of 60° with the horizontal, moves a body a distance of 10 m along the horizontal.

Qu. 3 (a) Calculate the efficiency of a machine supplied with a power input of 1 kW which produces a power output of 700 W.

(b) A machine is 60% efficient. Calculate the energy input to produce an output of 30 kJ.

Qu. 4 The power input to a machine of 80% efficiency is 100 W. Calculate:
(a) the power output;
(b) the energy output in a time of ten minutes;
(c) the energy losses in the same time of ten minutes.

Calculate:
(a) the power output from a machine of 70% efficiency when the input power is 300 W;
(b) the efficiency of a machine with an input power of 500 W which delivers an output power of 450 W;
(c) the energy input from a machine of efficiency 55% so as to produce an output of 100 kJ.

4 Name five forms of energy and describe two examples of processes by which one form of energy can be converted to another.

5 Plot the force–distance graph for the following data:

Distance (m)	0	10	20	30	40	50	60	70	80
Force (N)	0	13	27	40	54	60	60	60	60

and estimate the total work done over the distance 0 to 80 m.

6 A force F varies with distance s according to the formula $F = 100 - 5s$. Plot the $F - s$ graph from $s = 0$ to $s = 20$ m and calculate:
(a) the total work done over the distance 0 to 20 m;
(b) the work done over the interval $s = 5$ to $s = 15$ m.

Problems 8

1 Define the following terms and state their SI units:
(a) work; (b) energy; (c) power;
(d) efficiency.

2 Calculate the work done when:
(a) a force of 500 N moves a body of mass 100 kg a distance of 250 m in the direction of the force;
(b) a body of mass 62 kg is lifted through a height of 12.2 m (take $g = 9.81\ \text{m/s}^2$);
(c) a force of magnitude 80 N moves a body a distance of 50 m and the direction of the force makes an angle of 30° with the distance direction.

Define efficiency and explain why efficiency cannot exceed 100% for a practical machine.

9 Heat and temperature-change effects

General learning objectives: to solve problems associated with mass, specific heat capacity and temperature change, showing how materials expand or contract with temperature change and to illustrate positive and negative effects in practical situations.

9.1 Heat and temperature

Heat is a form of energy which may be converted into other forms of energy. The quantity of heat is therefore measured in units of energy, joules (J).

Temperature describes the degree of 'hotness' or 'coldness' of a body. It is a means of specifying the level of sensation caused by heat energy and *not* the quantity of heat energy, although, of course, a hot body at a high temperature will contain more heat energy than the same body at a lower temperature.

Temperature also defines the property which determines the direction of flow of heat energy. Heat energy always flows from a higher to a lower temperature, unless mechanical energy is expended to force it to do otherwise. For example, if a body at a higher temperature is immersed in water, heat from the body will be transferred to the water, with the result that the temperature of the water rises and the temperature of the body falls until equilibrium is established with the two temperatures equal. If this occurs under well-lagged conditions so that no heat escapes to the surroundings then equilibrium is established when the heat loss of the body equals the heat gain by the water.

In order to produce a temperature scale, a minimum of two fixed points must be selected to define the scale. The two fixed points of the Celsius scale are:

1 The ice point, the lower fixed point, which is defined as the temperature of melting ice, or to be absolutely specific, the temperature of equilibrium of ice, liquid water and its vapour at the standard atmospheric pressure of 1.01325×10^5 Pa.
2 The steam point, the upper fixed point, which is defined as the temperature of steam rising from pure water boiling under standard atmospheric pressure.

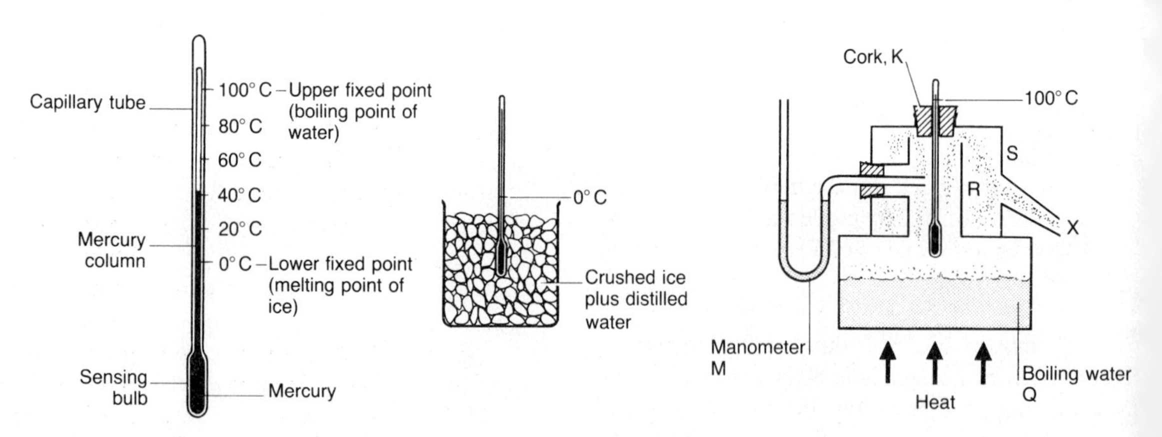

(a) Mercury in glass thermometer calibrated on Celsius scale

(b) Determination of lower fixed point on Celsius scale

(c) Determination of upper fixed point on Celsius scale

Figure 9.1

On the Celsius temperature scale, the ice point is taken as zero degrees Celsius, and written as 0°C. (To be absolutely rigorous, the ice point is internationally agreed to be 0.01°C, but for all but the most accurate measurement negligible error occurs in assuming the ice point to be 0°C.) The steam point is taken as 100 degrees Celsius and is written as 100°C. The temperature interval between the lower and upper fixed points is divided into 100 equal divisions. Figure 9.1 illustrates the temperature scale and how the lower and upper fixed points could be determined for the case of a mercury thermometer.

The SI unit of temperature is the kelvin (K). The kelvin is defined as 1/273.16 of the temperature of the ice point; this point is also known as the triple point of water. A temperature interval of one kelvin is exactly equal to a temperature interval of one degree Celsius. The kelvin temperature scale is the *absolute* temperature scale. The relationship between corresponding temperatures on the two scales are:

0 K = −273.15°C, absolute zero of temperature.
273.16 K = +0.01°C, the triple point of water (ice point).
373.15 K = +100°C, the steam point.

To convert a temperature quoted in degrees Celsius to kelvin (absolute temperature), add 273.15, usually 273 is accurate enough; to convert kelvin to degrees Celsius, subtract 273.15. For example:

$$0°\text{C} = 273.15\,\text{K}$$
$$20°\text{C} = 273.15 + 20 = 293.15\,\text{K}$$
$$-40°\text{C} = 273.15 - 40 = 233.15\,\text{K}$$
$$150\,\text{K} \approx 150 - 273 = -123°\text{C}$$
$$500\,\text{K} \approx 500 - 273 = 227°\text{C}$$

.2 Specific heat and problems ssociated with mass, specific heat apacity and temperature change

et us now put the ideas of heat and temperature hange involved for different substances on a uantitative basis by first defining specific heat apacity.

The **specific heat capacity** of a substance is the amount of heat energy transferred to change the temperature of one kilogram mass of the substance by one kelvin (or equivalently one degree Celsius).

The units of specific heat capacity are joules per kilogram per kelvin, written in symbol form as $\text{J kg}^{-1}\text{K}^{-1}$. Also since an interval of 1 K ≡ 1°C, the units may also be expressed as $\text{J kg}^{-1}°\text{C}^{-1}$. Specific heat capacities depend on temperature range. Values for some common substances are given in Table 9.1 and for ice and water in Table 9.2.

With the definition of specific heat capacity in mind, we are now in a position to solve heat problems associated with mass, specific heat capacity and temperature change. We use two basic facts, the first based on the definition of specific heat capacity, the second based on applying the law of conservation of energy.

Table 9.1 Specific heat capacities of some common substances

Substance	*Specific heat capacity c* ($\text{J kg}^{-1}\text{K}^{-1}$)	*Temperature range* (°C)
Aluminium	908	17–100
Copper	385	15–100
Mercury	139.3	20
Iron	460	18–100
Steel	450–480	15–100
Sand	800	20–100
Stone	750–960	0–30
Ethyl alcohol	2290	0

Table 9.2 Specific heat capacity for ice and water

	Specific heat capacity c ($\text{J kg}^{-1}\text{K}^{-1}$)	*Temperature* (°C)
Ice	2000–2090	−21–0
Water	4217	0
	4192	10
	4186	15
	4182	20
	4178	30–40
	4184	60
	4196	80
	4215	99

1 The amount of heat energy required to change the temperature of a mass of substance is given by,

$$H = mc(T_2 - T_1) \text{ joules}$$

where m = mass in kilograms (kg)
c = specific heat capacity, $\text{J kg}^{-1}\text{K}^{-1}$
$T_2 - T_1$ = temperature change in kelvin or degrees Celsius.

The amount of heat is given out when the substance cools and is absorbed when the substance is heated.

2 The total heat 'lost' from a cooling body or bodies equals the total heat 'gained' by initially cooler body or bodies.

For example, if a body of mass m_1, specific heat capacity c_1 and temperature T_1°C is placed in a container of liquid which is at a lower temperature T_2, and the masses of the container and liquid are m_2 and m_3 and their respective specific heat capacities are c_2 and c_3, then the final temperature T°C of all three bodies when equilibrium is established is found from applying

$$\text{Total heat lost} = \text{Total heat gained}$$

i.e. $m_1c_1(T_1 - T) = m_2c_2(T - T_2) + m_3c_3(T - T_2)$

Examples

1 Calculate the total heat energy required to raise the temperature of 100 kg of water contained in a copper tank of mass 20 kg from 12°C to 48°C. Neglect heat losses to the surroundings and take the specific heat capacities of water and copper as $4180\,\text{J kg}^{-1}\text{K}^{-1}$ and $385\,\text{J kg}^{-1}\text{K}^{-1}$ respectively. If the energy is supplied by an immersion heater delivering 3 kW of power determine the time taken for the heating process.

Solution

Heat required to raise m = 100 kg of water from 12°C to 48°C,

$$H_w = \text{mass} \times \text{specific heat capacity} \times \text{temperature rise} = 100 \times 4180 \times (48 - 12) = 15.048\,\text{MJ}$$

Heat must also be supplied to raise the temperature of the copper tank container,

$$H_c = mc(T_2 - T_1) = 20 \times 385 \times (48 - 12) = 0.2772\,\text{MJ}$$

so the total heat, $H = H_w + H_c = 15.048 + 0.2772 = 15.3252\,\text{MJ}$ *Ans*

If energy is supplied at the rate of 3 kW (3000 joules per second) then the time taken,

$$t = \frac{15.3252 \times 10^6}{3 \times 10^3} = 5108\,\text{s or } 1.42\,\text{h} \quad Ans$$

2 A 0.75 kg mass of iron at a temperature of 100°C is immersed in a beaker of water containing 1.5 kg of water at 10°C. The thermal capacity of the beaker is $120\,\text{J kg}^{-1}$. (Note: thermal capacity = mass × specific heat capacity). Assuming the beaker is well lagged so heat losses to the surroundings can be neglected, calculate the temperature to which the water rises. Specific heat capacity of iron and water are 460 and $4180\,\text{J kg}^{-1}\text{K}^{-1}$ respectively.

Solution

Let the temperature to which the water and beaker rise and that to which the iron mass falls be T°C, then heat transferred by iron in cooling from 100 to T°C is

$$H_i = mc(100 - T) = 0.75 \times 460 \times (100 - T) = 345(100 - T)$$

Heat gained by water and beaker in rising from 10 to T°C

$$H_w = m_wc_w(T - 10) + 120(T - 10)$$
$$H_w = [1.5 \times 4180 \times (T - 10)] + 120(T - 10) = 6270\,(T - 10) + 120(T - 10) \quad (2)$$

To find T we equate (1) and (2), i.e. heat lost = heat gained,

so $345(100 - T) = 6270(T - 10) + 120(T - 10)$

$34\,500 - 345T = 6270T - 62\,700 + 120T - 1200$

$6270T + 120T + 345T = 34\,500 + 62\,700 + 120$[illegible]

$$6735T = 98\,400$$

$$T = \frac{98\,400}{6735}$$

$$= 14.6°C \qquad Ans$$

9.3 Sensible and latent heat and temperature–time graphs for changes of state

Sensible heat is that heat whose effect is observed by an increase or decrease in temperature. For example, if a substance gains an amount H of heat energy, its temperature rises from T_1 to T_2 according to the equation

$$H = mc(T_2 - T_1)$$

where m is mass of the substance and c is its specific heat capacity. Likewise if a substance cools from a temperature T_3 to a lower temperature T_4, the heat energy given out is

$$H' = mc(T_3 - T_4)$$

In both cases, a temperature change is observed and H and H' are sensible heats.

When a substance changes state, that is, it changes from solid to liquid form, or from liquid to gaseous or vapour form, heat energy must be supplied to effect the change. While the change is taking place, there is normally no change of temperature. All the heat energy is used to effect the change of state. The heat energy required to change the state of a substance is known as **latent heat**.

We can normally define two latent heats for a substance: one known as the latent heat of fusion, which refers to the change of a substance from the solid to the liquid state and vice versa; and the second known as the latent heat of vaporization which refers to the change from the liquid to the gaseous state. In both cases, no change in temperature is observed when the substance is changing its state.

The term fusion is used to describe the melting of a substance which occurs at a specific temperature known as the melting or freezing point temperature (for example the freezing point of water is also the melting point of ice). When a substance is melted, heat must be supplied. When substance freezes, heat is given out. The term vaporization is used to describe the change of a substance from the solid or the liquid state to a gas or vapour. A vapour is a gas, but 'vapour' is normally used to describe specifically a gas of a substance which exists mainly as a solid or liquid at normal temperatures and atmospheric pressure, e.g. water vapour, petrol vapour, alcohol vapour, scent.

Some substances, e.g. naphthalene, change directly from the solid to the vapour state and in these cases, where the liquid phase is absent, we describe the vaporization as sublimation. When a substance is vaporized, heat must be supplied to effect the transition. When a vapour condenses back to liquid or solid, heat is evolved. During both transitions there is normally no change in temperature – hence, as in the case of fusion, latent heat is either absorbed or evolved without any change in temperature.

The distinction between sensible and latent heat is illustrated graphically in Figure 9.2 which shows the temperature–time graph for the transition of ice-to-water and water-to-steam for the case where heat energy is supplied at a uniform rate.

Point A, the initial point on the graph of Figure 9.2, refers to a quantity of ice at −20°C. Over the range AB, the heat supplied is exhibited as sensible heat being used to raise the temperature

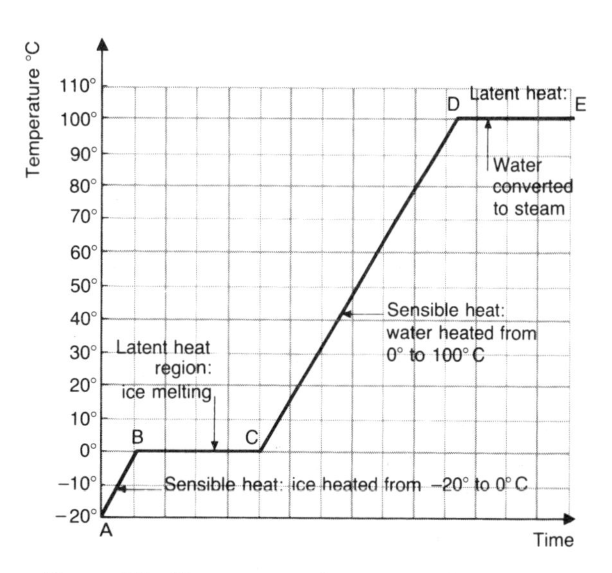

Figure 9.2 Temperature–time curve showing transitions of ice-to-water-to-steam

of the ice from −20°C to its melting point at 0°C. At point B, heat energy is now required to melt the ice to water. Over the range BC the temperature remains constant at 0°C until all the ice has melted, i.e. changed state from ice to water. The heat supplied to effect the change is latent heat. Subsequently, over range CD, sensible heat is absorbed and the water is raised from 0°C to 100°C, the boiling point of water. Latent heat (greatly in excess of the value required to melt the ice) must now be supplied to convert the water to steam. No change of temperature from 100°C will occur until all the water has been vaporized to steam. Thus, we obtain the flat range DE. Finally, although not shown on the graph, the steam absorbs sensible heat and its temperature rises. In this final range, we say we have superheated steam.

Figure 9.3 illustrates a second example of a temperature–time graph showing the sensible and latent heat region for a pure metal cooling from an initial liquid or molten state above its melting point. Sensible heat is given out over the range KL as the molten metal cools to its melting/fusion point temperature; latent heat is given out over the range LM as the metal solidifies and the temperature remains constant until all the metal has changed state from liquid to solid; finally sensible heat is given out over the region MN as the solid metal cools.

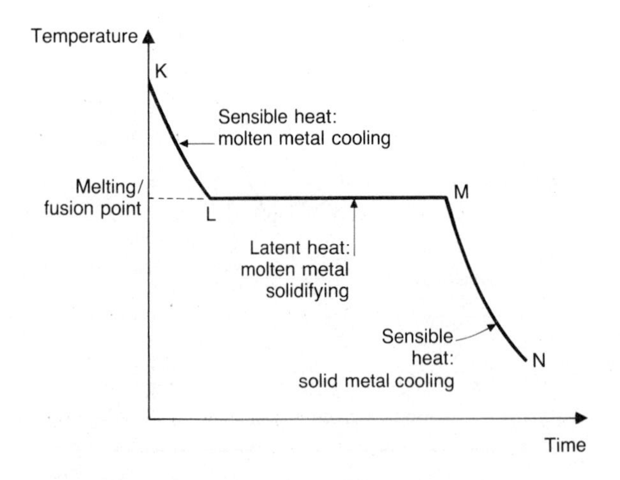

Figure 9.3 Temperature–time graph for a metal initially above its melting point cooling to solid state

9.4 Specific latent heat and change-of-state problems

The specific latent heat of fusion of a substance is the amount of heat energy required to completely change one kilogram of the substance from the solid to the liquid state at its melting point temperature, without change in its temperature or change in pressure.

The specific latent heat of vaporization of a substance is the amount of heat energy required to completely change one kilogram of the substance from the liquid to the gaseous (or vapour) state at its boiling-point temperature, without change in its temperature or change in pressure.

In cases where sublimation occurs (i.e. direct change from solid to gaseous state), replace 'liquid' by 'solid' in the above definition.

Specific latent heats are normally denoted by the symbol L and the units of the specific latent heat are joules per kilogram, J/kg or J kg^{-1}. Table 9.3 gives values for the specific latent heat for some common substances.

Thus the total heat energy required to change the state of a substance from solid to liquid at its

Table 9.3 Specific latent heat values

Substance	*Melting point* (°C)	*Latent heat of fusion* (J/kg)
Aluminium	660.1	400×10^3
Copper	1083	200×10^3
Gold	1063	67×10^3
Silver	960.8	105×10^3
Iron	1535	210×10^3
Sulphur	112.8	38×10^3
Mercury	−38.9	12×10^3
Ice	0	333.5×10^3
	Boiling point (°C)	*Latent heat of vaporization* (J/kg)
Water	100	2.2564×10^6
Mercury	356.6	0.276×10^6
Ether	34.6	0.375×10^6

melting point or from liquid to gas at its boiling point is given by

$H_L = mL$ joules

where m = mass of substance in kilograms

L = specific latent heat in joules per kilogram (note the symbol h is also used to denote specific latent heat).

An equal amount of heat is, of course, given out when the reverse process occurs, i.e. when a vapour converts to a liquid or a liquid converts to a solid.

Examples

1 An electrical heater supplies 3 kW of power to melt ice at 0°C. Calculate the mass of ice melted per hour, given the specific latent heat of fusion of ice is 333.5 kJ/kg.

Solution

Let m = mass of ice melted per hour, then the energy supplied by the heater,

$$\begin{aligned} H &= \text{power} \times \text{time} \\ &= 3000\,\text{W} \times (1\ \text{hour}) \\ &= 3000 \times 60 \times 60\,\text{J} \\ &= m \times L, \text{ where } L = 333\,500\,\text{J/kg} \end{aligned}$$

Thus $m = \dfrac{3000 \times 60 \times 60}{333\,500} = 32.38\,\text{kg}$ *Ans*

2 Calculate the amount of heat required to melt 20 kg of aluminium at an initial temperature of 20°C, given the following information:

Specific heat capacity of aluminium $= 908\,\text{J kg}^{-1}\,\text{K}^{-1}$;
melting point of aluminium = 660°C;
specific latent heat of aluminium = 400 kJ/kg.

Solution

Heat required = sensible heat to raise from 20 to 660°C + latent heat to melt aluminium

$$\begin{aligned} \text{Sensible heat} &= mc(T_2 - T_1) \\ &= 20 \times 908 \times (660 - 20) \\ &= 11.622\,\text{MJ} \end{aligned}$$

$$\begin{aligned} \text{Latent heat} &= mL \\ &= 20 \times 400\,\text{kJ} = 8\,\text{MJ} \end{aligned}$$

so total heat required $= 11.622 + 8$
$= 19.622\,\text{MJ}$ *Ans*

9.5 The effect of temperature change on the physical dimensions of materials: coefficients of expansion

Most substances, whether in the solid, liquid, or gaseous state, expand when heated, that is as their temperature rises, and contract on being cooled. One notable exception to this rule is water, which actually contracts in volume as it is heated from 0°C to 4°C. However, over the range 4°C to 100°C, it expands with increasing temperature. Expansion and contraction occur in all three dimensions, but for solid materials we may quantify the effect of expansion in three ways:

1 Linear expansion, which refers to the changes in length with temperature, and is of practical importance when dealing with wires, rods, cables, rails, etc.
2 Area expansion, which refers to changes in area with temperature, and is useful when dealing with thin sheets of materials.
3 Volume expansion which refers to changes in volume with temperature.

Since liquids and gases cannot 'hold' their shape, we can only really quantify expansion effects as a change in volume with temperature, although in the case of a gas, we must also consider pressure.

It is found, experimentally, that the length of many solid materials varies with temperature over a given temperature range according to the formula:

$$l_2 = l_1[1 + \alpha(T_2 - T_1)]$$

where l_2 = length at temperature T_2

l_1 = length at temperature T_1

α = a constant (Greek letter alpha), known as the *coefficient of linear expansion*

Using this formula, we have

$$l_2 = l_1 + \alpha l_1(T_2 - T_1)$$

so change in length for a temperature change of $T_2 - T_1$

$$l_2 - l_1 = \alpha l_1(T_2 - T_1)$$

i.e. expansion or contraction = α × original length × temperature change.

and making α the subject of the formula, we obtain

$$\alpha = \frac{l_2 - l_1}{l_1(T_2 - T_1)}$$

So α, the coefficient of linear expansion, may be defined as the change in length divided by the original length per degree Celsius (or kelvin); α has the 'units' of per °C or per K, $°C^{-1}$ or K^{-1}. It is usual practice to take l_1 as the reference length at temperature $T_1 = 20°C$ although $T_1 = 0°C$ is also often used.

The volume expansion of solids, liquids, and gases (at constant pressure) is given by a very similar formula:

$$v_2 = v_1[1 + \gamma(T_2 - T_1)]$$

where v_2 = volume at T_2, v_1 = volume at T_1 and γ (Greek letter gamma) is a constant for a given substance. γ is known as the coefficient of volume expansion.

The expansion constants vary slightly with temperature and it is usual practice to quote α and γ as the average value over a given temperature range. Table 9.4 gives values of α and γ for some common substances at 20°C.

Table 9.4 Coefficients of linear and volume expansion of some substances at 20°C

Substance	*Coefficient of linear expansion, α* (K^{-1} or $°C^{-1}$)	*Coefficient of of volume expansion, γ* (K^{-1} or $°C^{-1}$)
Aluminium	24×10^{-6}	72×10^{-6}
Brass (60%Cu, 40%Zn)	20×10^{-6}	60×10^{-6}
Copper	17×10^{-6}	51×10^{-6}
Steel	12×10^{-6}	36×10^{-6}
Glass	$7\text{–}10 \times 10^{-6}$	$20\text{–}30 \times 10^{-6}$
Wood		
(along grain)	$3\text{–}5 \times 10^{-6}$	$70\text{–}130 \times 10^{-6}$
(across grain)	$35\text{–}60 \times 10^{-6}$	
Invar-steel alloy	2×10^{-6}	6×10^{-6}

Examples

1 Calculate the change in length of an aluminium bar of length 250 mm at 20°C when its temperature is raised to 100°C. The coefficient of linear expansion of aluminium is:

$24 \times 10^{-6}\,°C^{-1}$.

Solution

$$\begin{aligned}\text{Change of length} &= \alpha \times \text{original length} \\ &\quad \times \text{temperature change} \\ &= 24 \times 10^{-6} \times 250 \\ &\quad \times (100 - 20) \\ &= 480\,000 \times 10^{-6} \\ &= 0.48\,\text{mm} \quad \textit{Ans}\end{aligned}$$

2 Steel cables used to support a suspension bridge are nominally 1000 m in length. Climatic conditions are such that the bridge is unlikely to experience temperatures exceeding +42°C in summer nor fall below −10°C in winter. Calculate the variation in cable length between these two extremes. The coefficient of linear expansion for steel, $\alpha = 12 \times 10^{-6}\,°C^{-1}$

Solution

Maximum variation in length

$= \alpha \times 1000 \times (T_2 - T_1)$

where $T_2 = 42°C$ and $T_1 = -10°C$, hence

$$\begin{aligned}\text{length variation} &= 12 \times 10^{-6} \times 1000 \\ &\quad [42 - (-10)] \\ &= 12 \times 10^{-6} \times 1000 \times 52 \\ &= 0.624\,\text{m} \quad \textit{Ans}\end{aligned}$$

3 The length of a metal rod is accurately measured at 20°C and is found to be 192.40 mm. The rod is then heated to 60°C and its expansion is measured as 0.35 mm. Calculate the coefficient of linear expansion of the metal.

Solution

The coefficient of linear expansion,

$$\alpha = \frac{\text{change in length}}{\text{length at 20°C} \times \text{temperature change}}$$

$$= \frac{0.35}{192.40 \times (60 - 20)}$$

$$= 45.5 \times 10^{-6}\,°C^{-1} \quad \textit{Ans}$$

9.6 Practical examples and design implications of thermal expansion and contraction

Thermal expansion and contraction effects may be utilized in a wide variety of practical applications. The expansion of mercury is used in the mercury thermometer; the expansion of a wire when heated by electrical current is used in the hot-wire ammeter to measure current, and it is also applied in light-flashing units; steel plates may be joined together by red-hot rivets, which contract on cooling, thus holding the plates securely together; a red-hot metallic sleeve may be slid over a cylinder and, on cooling, has a high-pressure bond to the cylinder. Alternatively, the cylinder may be cooled by refrigeration, and then the sleeve placed in position; a 'stuck' jam-jar lid may be released by heating the lid under the hot tap. Practical examples of differential expansion effects are given below.

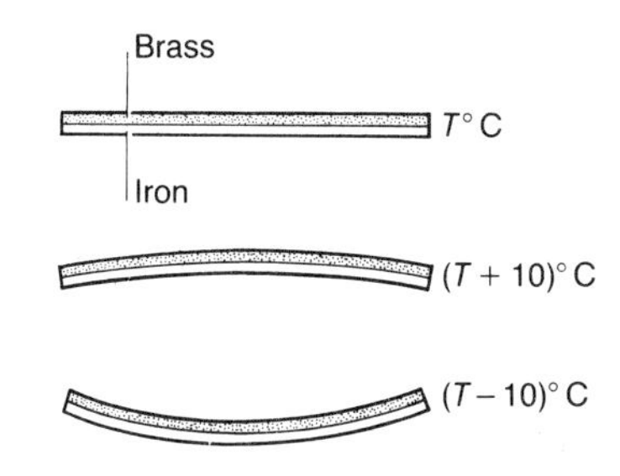

(a) Action of bimetallic strip with temperature change

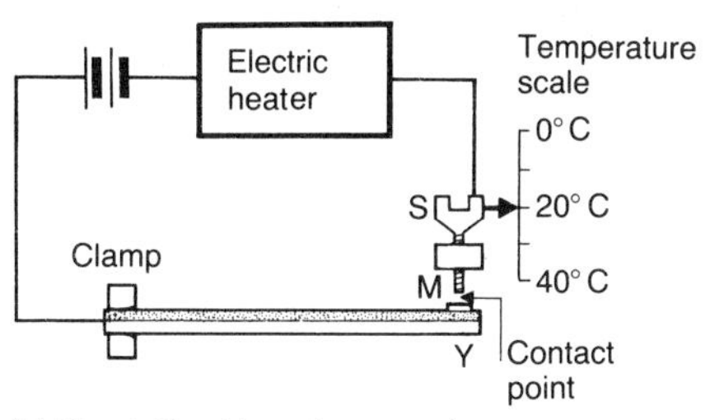

(b) Bimetallic strip acting as a thermostat

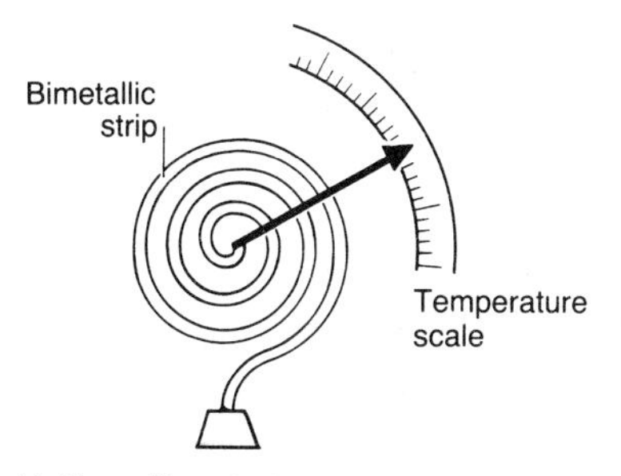

(c) Bimetallic strip thermometer

'igure 9.4 Applications of the bimetallic strip

If two metallic strips of different expansion coefficient are either bonded or riveted together, then when heat is applied, the difference in the expansions of the two materials will cause the compound strip, usually referred to as a bimetallic strip, to bend. For example, the bimetallic brass and iron strip shown in Figure 9.4(a) will be straight at a given temperature. When the temperature is raised, it will bend with the brass material on the outside of the curve, since brass has a greater (about 1.7 times) coefficient of linear expansion. If the temperature is lowered below the 'straight' value. It will bend the other way, since brass contracts more than iron.

Use is made of bimetallic strip elements in thermostats in controlling heating and cooling processes. For example, in Figure 9.4(b), electric power is supplied to an appliance by a battery via the bimetallic strip when the unclamped end Y is in contact with the small magnet M (M gives a positive snap action on subsequent closing and cuts down sparking). When the temperature rises, the bimetallic strip bends and opens the contact, thus cutting off the current to the appliance. As the temperature falls, the bimetallic strip unbends and, when the temperature reaches its original temperature the strip, returns to its original position and closes the contact, thus re-making the circuit and allowing current to flow. The temperature at which the circuit 'breaks' can be adjusted by varying the position of M by means of screw S. If a relatively long bimetallic strip is coiled into a flat spiral, and one end of the coil is clamped, the other end shows a considerable movement as temperature changes. This principle is utilized in the measurement of temperature. A bimetallic strip thermometer is shown in Figure 9.4(c).

The effect of thermal movement must constantly be taken into account in industrial design. Steel rails must be laid with gaps between them, otherwise a temperature rise would cause severe buckling due to the tremendous forces involved when the steel rail tends to expand. However, nowadays steel railway lines are continuously welded together and therefore any expansion/contraction effects taking place must be taken up by the materials being able to withstand the large compressive/tensile stresses that may result. In an engine, moving parts, such as the pistons in the cylinder block and the associated timing and valve components, must be designed to allow for expansion under operating conditions. Pendulums and balance wheels of watches depend on accurate dimensions and therefore expansion and contraction effects must be minimized to avoid time errors. This is accomplished normally by using either specially developed alloy materials which have a very low expansion coefficient, or compensation techniques where the expansion of one component counteracts the expansion of another.

In bridge and building construction, expansion joints must be provided. In accurate measurement using steel tapes, micrometers, etc., the temperature of measurement must be noted and corrections made for expansion or contraction in these instruments, if measurements are taken at a temperature different from the temperature of their calibration. Measuring instruments are normally calibrated at 20°C.

Test 9

This test may be used as a basic self-assessment test to check whether you have absorbed the main facts of Chapter 9 on **Heat and temperature change effects**, and its learning objectives. All answers to be entered in the answer block.

Qu. 1 Enter a tick (√) in the answer block if you consider the statement is correct; enter a cross (×) if you consider the statement is in any way incorrect.

(a) The specific heat capacity of a substance is the amount of heat energy required to change the temperature of 1 kg of the substance by 1°C.

(b) Latent heat is the amount of heat required to change the state of 1 kg of substance, e.g. from solid to liquid, from liquid to vapour.

(c) The specific latent heat of fusion of a substance equals the specific heat capacity of the same substance.

(d) If brass has a coefficient of linear expansion equal to $20 \times 10^{-6}\,K^{-1}$, the increase in length of a one-metre bar over the temperature range 0–100°C is 2 cm.

Qu. 2 Calculate the amount of heat required to raise the temperature of 5 kg of water from 20°C to 85°C. Neglect any heat losses and take the specific heat capacity of water as $4.2\,kJ\,kg^{-1}\,K^{-1}$.

Qu. 3 A 0.5 kg mass of copper at 100°C is immersed in 1 kg of water at 20°C. Assuming no heat losses, and taking the specific heat capacities of copper and water as $385\,J\,kg^{-1}\,°C^{-1}$ and $4200\,J\,kg^{-1}\,°C^{-1}$ respectively, calculate the temperature to which the water rises.

Qu. 4 The length of a steel beam measured at 0°C is 10 m. If the coefficient of linear expansion of steel is $12 \times 10^{-6}\,°C^{-1}$, determine

(a) the amount of contraction if the temperature is lowered to −20°C;

(b) the temperature at which the 10 m length is increased by 10 mm.

Problems 9

1 Distinguish between the following:
(a) heat and temperature;
(b) sensible and latent heat;

Answer block:

Question no.	1				2	3	4	
	(a)	(b)	(c)	(d)			(a)	(b)
Answer								

(c) degrees Celsius and absolute temperature measured in kelvin.

Define the specific heat capacity of a substance. Calculate the quantity of heat required to raise the temperature of 500 kg of water contained in a copper tank of mass 50 kg from 10°C to 50°C. Assume the tank is well lagged so any heat losses to the surroundings can be neglected and take the specific heat capacities of water and copper as $4186\,J\,kg^{-1}°C^{-1}$ and $385\,J\,kg^{-1}°C^{-1}$, respectively.
If the heat is supplied by a 3 kW heater, estimate the time for the rise to take place.
A 0.4 kg mass of metal at 100°C is immersed in 0.5 kg of water at an initial temperature of 17°C. It is observed that the temperature of the water rises to 23.6°C. If heat losses to the surroundings can be neglected, calculate the specific heat capacity of the metal.

Sketch the characteristic temperature–time graph for the cooling of a metal initially in its molten state.
5 kg of aluminium in its molten state and at its melting point of 660°C is cooling to its solid state. What total heat is given out in the change-of-state process? The specific latent heat of fusion of aluminium is $400\,kJ\,kg^{-1}$.

10 kg of ice at −10°C is heated by a 3 kW source of power. Calculate, assuming no external heat losses:
(a) the heat required to raise the ice to 0°C;
(b) the heat required to melt the ice to water at 0°C;
(c) the heat required to raise the temperature of the water from 0 to 100°C;
(d) the heat required to convert the water at 100°C to steam at 100°C;
(e) the total time for the complete processes (a) to (d).
Take the specific heat capacities of ice and water as $2000\,J\,kg^{-1}K^{-1}$ and $4180\,J\,kg^{-1}K^{-1}$; the latent heat of fusion of ice at 0°C as $334 \times 10^3\,J\,kg^{-1}$ and the latent heat of vaporization of water at 100°C as $2.26 \times 10^6\,J\,kg^{-1}$.
Calculate the mass of steam at 100°C required to raise the temperature of 250 kg of water contained in a copper tank of mass 40 kg from an initial temperature of 12°C to a final temperature of 55°C. The following data is given:
Specific heat capacities of copper and water are respectively 385 and $4180\,J\,kg^{-1}°C^{-1}$; latent heat of vaporization of water at 100°C = $2.26 \times 10^6\,J\,kg^{-1}$.

7 Describe two examples of
(a) the practical application; (b) the design implications, of the expansion/contraction of materials with temperature.

8 The length of a steel beam at T°C is given by

$$l = 5[1 + 12 \times 10^{-6}(T - 20)]$$

Calculate the length at $T = -20$°C, 0°C and 40°C.

It is known that the compressive force set up in the beam when its length is compressed by x metres is given by

$$F = 40 \times 10^6 \times x \text{ newtons}$$

Calculate the compressive force set up in the beam when it is clamped at each end at 20°C and its temperature is subsequently raised to 50°C.

9 Define the coefficient of linear expansion and suggest how you might measure its value for brass over the range 0 to 100°C. A brass bar has a length of 305.0 mm at 0°C and when it is heated to a temperature of 100°C its length increased by 0.61 mm. Calculate the coefficient of linear expansion for brass.

Part Four: Electricity

10 Introduction to electricity: electrical current and voltage

General learning objectives: to introduce standard circuit symbols for commonly used components in electrical circuits, to define the fundamental quantities of current and voltage and know their units, to understand the difference between direct current (dc) and alternating current (ac) and know examples of their use.

10.1 Introduction to electrical circuits and standard symbols for electrical components

In the study of electricity and electrical circuits we are concerned principally with two fundamental quantities: electrical current and voltage. Electrical current is the flow of charge in conductors and devices. Voltage is a measure of the 'pressure difference' set up by a source of electrical energy, such as a battery or electromagnetic generator, which acts to move charge and thereby produce current. We shall also be considering components which exhibit the property of restricting current and converting electrical energy to heat: the property of resistance and its relationship with voltage and current.

However, before we consider these important quantities in more detail let us first introduce the idea of a simple electrical circuit and the symbols used to represent electrical components on circuit diagrams.

Figure 10.1(a) shows a sketch of a simple electrical circuit consisting of a battery whose terminals are joined by a length of resistive wire. The battery acts as a source of energy – a voltage source able to generate current – and the resistive wire, the conducting medium allowing current flow from the positive (+) battery terminal to the negative (–) terminal.

In order to simplify having to draw a 'picture' of the components making up a circuit, we use internationally recognized symbols to denote the components. The circuit diagram representing Figure 10.1(a) is shown in (b). A fuller list of component symbols is given in Figure 10.2. A comprehensive list of symbols for electrical and

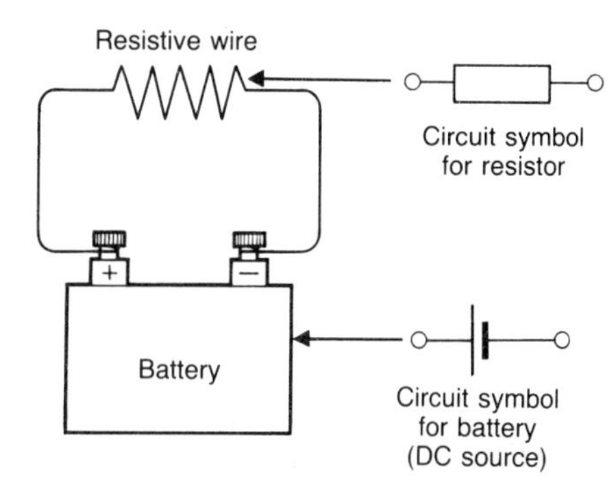

(a) A simple electrical circuit

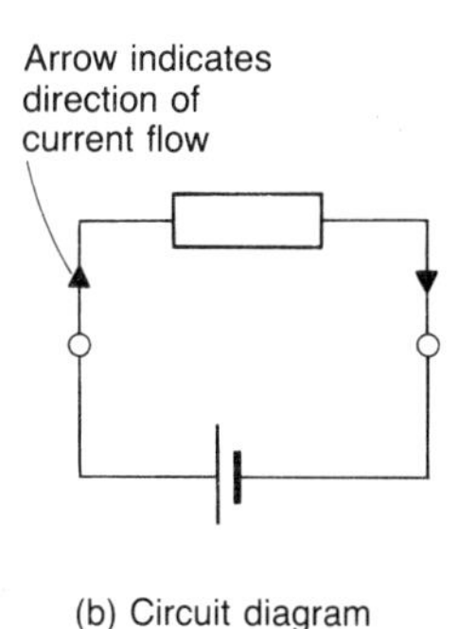

(b) Circuit diagram

Figure 10.1

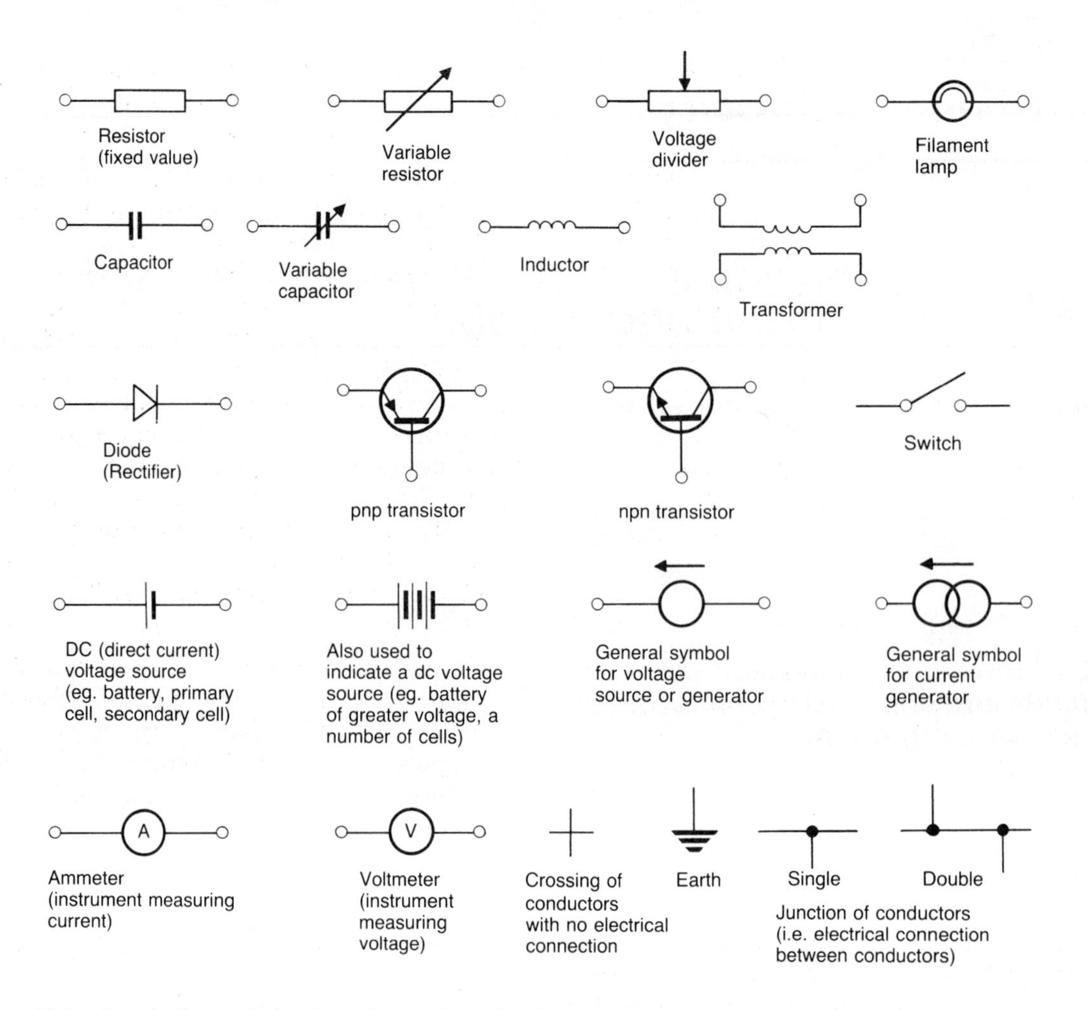

Figure 10.2 Standard symbols for electrical and electronic components

electronic components is given in the British Standard BS 3939 publication.

10.2 Electrical current: direct and alternating (dc and ac)

Electrical current may be defined as the rate of flow of charge. If we are dealing with a constant rate, then the current, normally denoted by the letter I is given by

$$I = \frac{\text{charge}}{\text{time}} = \frac{Q}{t}$$

where charge Q = total charge moving through the cross-section of the conducting medium,

t = time in seconds.

The SI unit of charge is the coulomb (C) and the unit of current is the ampere (A):

$1\,\text{C} = 1\,\text{As}; \qquad 1\,\text{A} = 1\,\text{C/s}$

Commonly used sub-multiple units of current are

the milliampere: $1\,\text{mA} = 0.001\,\text{A}$ or $10^{-3}\,\text{A}$
the microampere: $1\,\mu\text{A} = 10^{-6}\,\text{A}$

Currents of constant magnitude whose values do not vary with time and do not change direction are known as **direct currents**, normally abbreviated to **dc**. Voltage sources which generate direct current are also qualified in the same manner. Dc sources include dry batteries (used, for example, to power electronic calculators, torches, sma

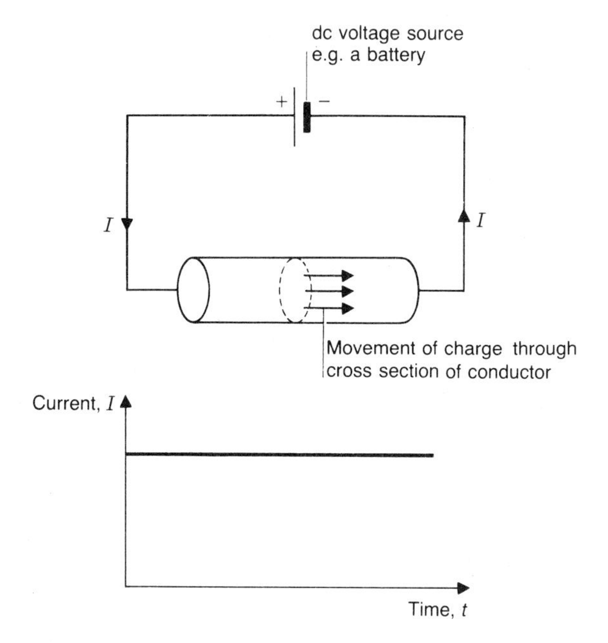

Figure 10.3 Direct current (dc), I *total charge moving through the cross-section of the conductor per second*

electric motors, digital watches, transistor radios); solar cells (used to convert the sun's energy to electrical energy and applied as an electrical source of power in an increasing number of applications from charging batteries to acting as the power source for portable electronic equipment); car batteries and electro-mechanical generators (the latter designed to produce dc for higher-power applications such as welding or to drive powerful dc motors). The majority of electronic equipment requires dc sources (even if this is derived from ac) for its operation.

Figure 10.3 shows a simple circuit producing dc current and the resulting waveform of current against time. The dc current remains constant both in magnitude and in the direction it flows.

Many sources, however, generate currents which vary with time. In such case we can define the instantaneous current,

$$i = \frac{\delta Q}{\delta t} \text{ amperes or coulombs per second}$$

where

δQ = quantity of charge moving through the cross-section of the conductor in a time interval δt,

δt = time interval between times t and $t + \delta t$, which is small compared with the overall time variation of the current.

An extremely important form of time-varying current is **alternating current** which varies in the form of a sinewave as shown in Figure 10.4. Alternating current, normally abbreviated to **ac**, is generated in a circuit when the voltage source varies in time in both its magnitude and its direction and therefore causes a current to flow in a circuit which varies continuously in magnitude but also actually changes its direction of flow.

In the late nineteenth century, and in the early twentieth century, the chief source of electrical energy supply was in the form of direct current electricity. However, today, virtually all countries in the world generate and distribute electrical energy in alternating current form. The main reason for this is that alternating current electricity may be generated and then subsequently transmitted over long distances very much more efficiently and economically than direct current.

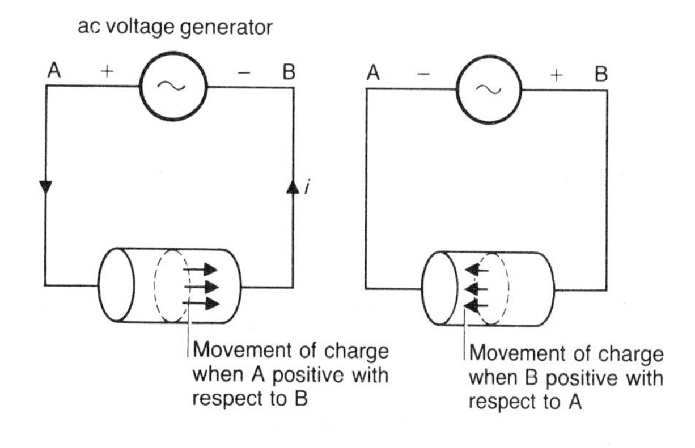

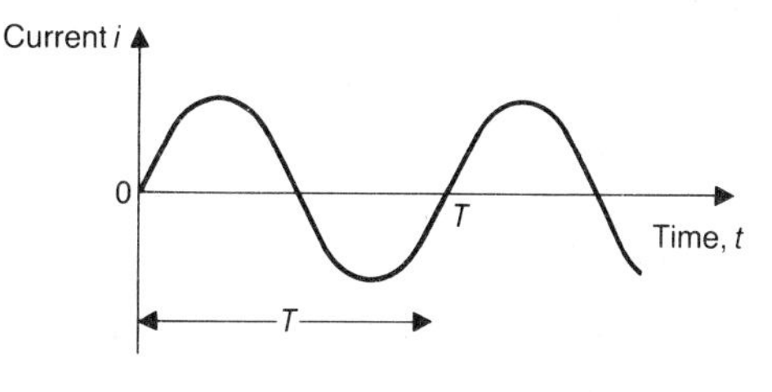

Figure 10.4 Alternating current (ac), i = *charge* δQ *moving through the cross-section of the conductor in a 'small' time* δt, $\delta t \ll T$ *where* T = *periodic time (time of one complete cycle). Note that* i *varies continually in magnitude and reverses its direction of flow every half-cycle*

Thus, the electrical mains supplies transmitted from the electrical power generating stations to our homes are ac electricity. The actual form of the voltage developed across the main supply terminals is of sinewave form, and when we plug an appliance into the mains supply, the ac voltage causes a sinewave ac current to flow in the leads to the appliance. An appliance, such as an electric fire, will therefore conduct an alternating current. The current flow in the leads and the heater elements will vary continuously in strength and will change its direction once every cycle. The waveform of the current is of sinusoidal form. Other appliances – e.g., a transistor radio or a battery charger – may require dc voltages and currents for their operation. In these cases, the ac supply from the mains must first be rectified, i.e. the ac mains input must be converted into a dc supply, and rectifier circuits will be incorporated in the appliance to effect this conversion. Likewise, electrical supplies to industry are of ac sinewave form, although the actual magnitude of the ac voltage may be higher than the values used for domestic purposes. AC mains supplies may be stepped up or down by means of transformers to provide a given ac output voltage. For the many tasks in industry which require dc supplies, rectifiers are used. Some examples where dc supplies are required are electroplating, electro-refining of metals, in dc motors and in dc electric railways.

The ac mains frequency is 50 Hz (50 cycles per second) for the United Kingdom and most other countries, with the notable exception of the United States where it is 60 Hz.

Examples

1 Figure 10.5 shows the waveform of an ac current. Determine

(a) the periodic time and frequency of the current;
(b) the peak amplitude of the current.

Solution

(a) The periodic time T can be found by measuring the time interval for one complete cycle, say between positive peaks. Hence

$$T = 20\,\text{ms or } 0.020\,\text{s} \quad \textit{Ans}$$

The frequency of the ac current (number of cycles per second),

$$f = \frac{1}{T} = \frac{1}{0.020} = 50\,\text{Hz} \quad \textit{Ans}$$

(b) The peak amplitude of the current is the maximum value of the current from zero in either positive or negative directions. So

$$\text{peak amplitude} = 2\,\text{A} \quad \textit{Ans}$$

2 The current flow in metallic conductors, see Figure 10.6, is carried by 'free' electrons each of which has a negative charge of magnitude 1.6×10^{-19} C. When acted upon by an electrical force the electrons have a uniform movement superimposed on their previous random paths. This movement is the actual current.
Determine how many electrons pass any given cross-sectional plane in a conductor when a current of 1 A is flowing.

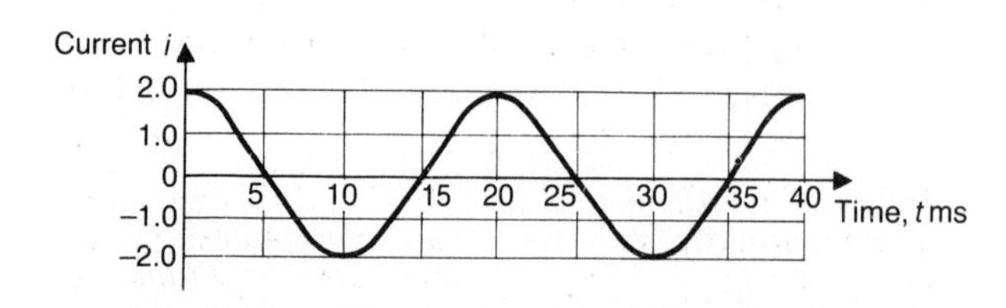

Figure 10.5 AC current waveform

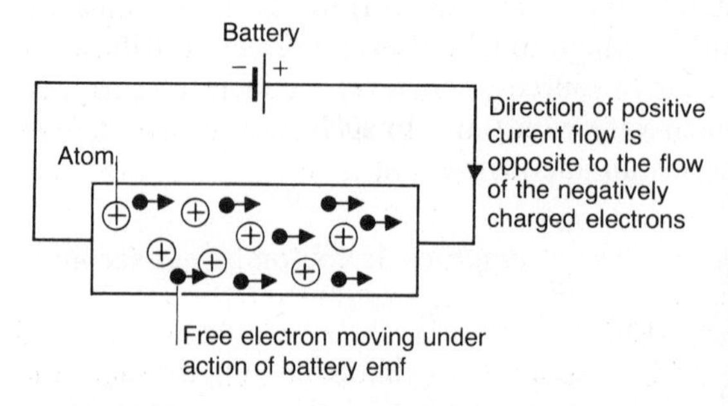

Figure 10.6

Solution

$$1 \text{ ampere} = 1 \text{ coulomb per second}$$
$$= \frac{1}{\text{electronic charge}}$$
$$= \frac{1}{1.6 \times 10^{-19}} = 6.25 \times 10^{18} \text{ electrons per second}$$

10.3 Voltage, electromotive force and potential difference

Current will not flow in a circuit unless a source of energy is applied to create a 'force' and produce a resultant flow of charge through the conductors and components comprising the circuit. The measure of the ability of a source to effect current flow in a circuit is defined by its **voltage** or, more descriptively, by the term **electromotive force (e.m.f.)**.

The voltage or e.m.f. of a source can be formally defined as the energy supplied by the source in transporting 1 coulomb of charge completely around a circuit or equivalently the power drawn from a source in sustaining a current of 1 ampere in a circuit.

The SI unit of voltage is the volt (V),

1 V = 1 J/C = 1 W/A
(J = joule, unit of work and energy; W = watt, unit of power; A = ampere, unit of current; C = coulomb, unit of charge.)

Multiple and sub-multiple units of the volt which are frequently used are:

kilovolts, 1 kV = 1000 V
millivolts, 1 mV = 0.001 V
microvolts, 1 μV = 10^{-6} V

An electrical source supplies energy to a circuit as soon as current flows in the circuit. For example, in the circuit of Figure 10.7 no energy is supplied and no current flows until switch S is closed. When the switch is open the conduction path for current is broken since current does not normally flow through air, air being a good insulator. When the switch is closed and the circuit is therefore 'completed', the source e.m.f. V_s drives current around the circuit. The source may be thought of as acting as a pump, pumping current through the circuit components in an analogous way that a water pump forces water around a central heating circuit.

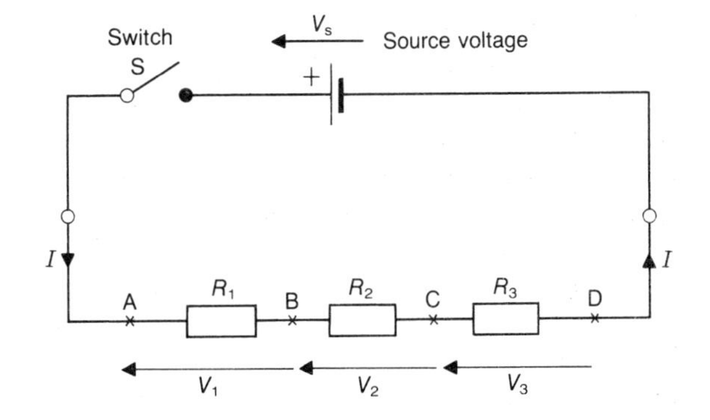

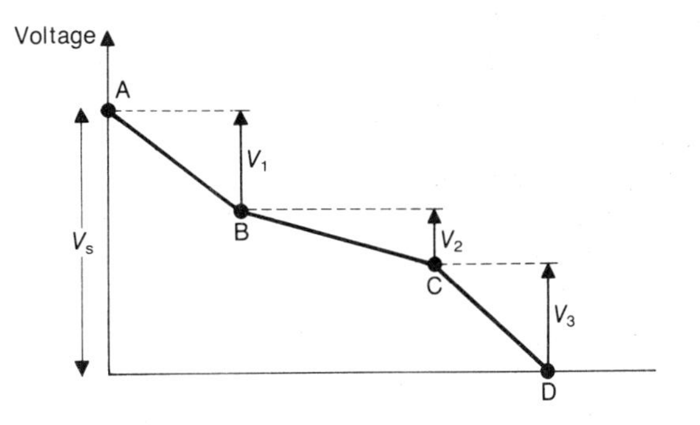

Figure 10.7 Source e.m.f., V_s = sum of voltages across components, i.e. $V_1 + V_2 + V_3$

As the current flows through each component in a circuit it dissipates energy and a potential difference or voltage is developed across each component, again analogous to the pressure difference which occurs between any two points in a pumped water circuit. The terms voltage, voltage or potential difference and voltage or potential drop are all used (and have identical meaning) to describe the difference in voltage levels that must be produced across a component to effect current flow.

The **voltage** or **potential difference** between any two points in an electrical circuit is defined as the work done in transporting unit charge (1 coulomb) between the two points. The energy for the work to be done is, of course, supplied by the voltage source. Thus if we apply the law of conservation of energy to the circuit of Figure 10.7 we have the very important result

applied e.m.f. =
sum of voltages across individual components

i.e. $V_s = V_1 + V_2 + V_3$

where V_1 = potential difference between points A and B
= voltage developed across component R_1, volts;

V_2 = potential difference between points B and C
= voltage developed across component R_2, volts;

V_3 = potential difference between points C and D
= voltage developed across component R_3, volts;

V_s = source e.m.f., volts;

In circuit diagrams, the direction of current flow is indicated by an arrow, the arrow normally being marked on one of the connecting 'wires' of the circuit. Current always flows from a higher voltage point to a lower voltage point. The potential difference between two points is also indicated by an arrow line, the point at the higher voltage level being denoted by the arrow tip as shown in Figure 10.7.

Test 10

This test may be used as a basic self-assessment test to check whether you have absorbed the main facts of Chapter 10 on **Introduction to electricity: electrical current and voltage**. All answers to be entered in the answer block.

Qu. 1 Enter a tick (√) in the answer block if you consider the statement is correct; enter a cross (×) if you consider the statement is in any way incorrect.
(a) Current is the rate of flow of charge and has the units of amperes.
(b) Direct current (dc) is constant in magnitude and direction while alternating current (ac) may vary both in magnitude and direction.

Answer block:

<table>
<tr><td rowspan="2">Question no.</td><td colspan="5">1</td><td colspan="3">2</td></tr>
<tr><td>(a)</td><td>(b)</td><td>(c)</td><td>(d)</td><td>(e)</td><td>(a)</td><td>(b)</td><td>(c)</td></tr>
<tr><td></td><td></td><td></td><td></td><td></td><td></td><td></td><td></td><td></td></tr>
<tr><td rowspan="5">Question 3</td><td></td><td colspan="7">Answers</td></tr>
<tr><td>(a)</td><td colspan="7"></td></tr>
<tr><td>(b)</td><td colspan="7"></td></tr>
<tr><td>(c)</td><td colspan="7"></td></tr>
<tr><td>(d)</td><td colspan="7"></td></tr>
<tr><td>Question 4</td><td colspan="8"></td></tr>
</table>

(c) The frequency of the ac mains supply in the UK and Europe is 50 Hz.
(d) The e.m.f. of a source is the energy supplied by the source in transporting unit charge (1 coulomb) completely around a circuit connected to the source terminals.
(e) The voltage between any two points in a circuit is the work done in transporting unit charge between the two points.

Qu. 2 Figure 10.8 shows the diagram of an ac waveform. Determine:
(a) the peak voltage amplitude of the waveform;
(b) the time interval between a voltage maximum and a voltage minimum (peak negative);
(c) the frequency of the ac waveform.

Qu. 3 Write down
(a) Two sources of dc energy;
(b) One source of ac energy;
(c) Two examples of the use of dc;
(d) An advantage of ac over dc.

Qu. 4 Using standard circuit symbols draw the circuit diagram for a dc battery source in series with a switch, a resistor and a lamp.

Problems 10

1 (a) Draw the standard circuit symbols for the following components: a dc source, an ac source, a resistor, a switch, a filament lamp.
(b) Draw the circuit diagram for a battery, a switch and three resistors joined in series.

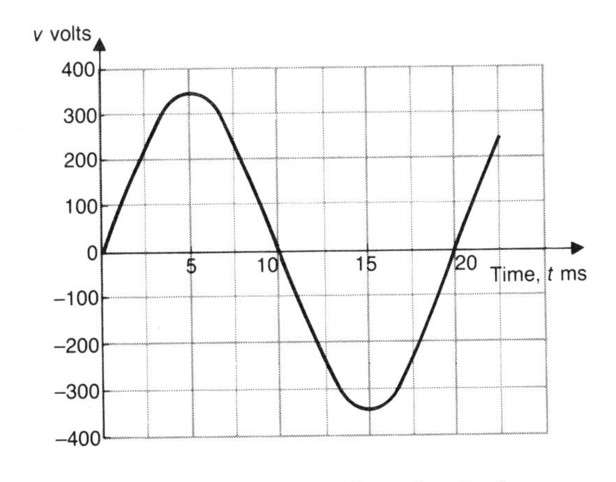

Figure 10.8 AC voltage waveform for Qu 2

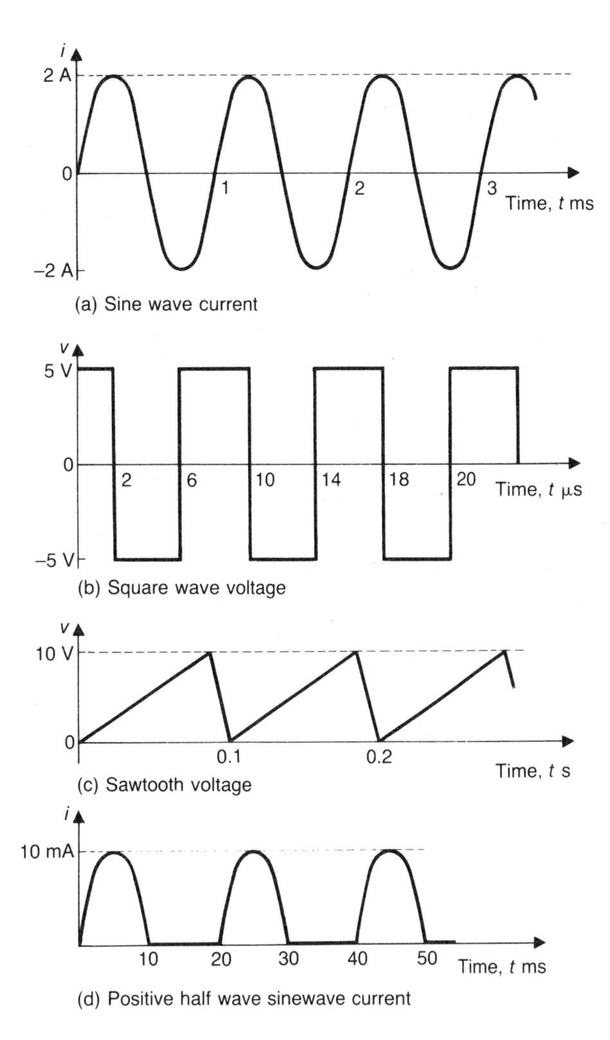

Figure 10.9 Waveforms for problem 3

2 Define the following quantities and state their units:
(a) current; (b) electromotive force (e.m.f.); (c) potential difference.

3 Draw a sketch of a sinusoidal ac current and mark on the sketch the peak amplitude and periodic time.
Determine for the periodic waveforms shown in Figure 10.9 the peak amplitude and repetition frequency.

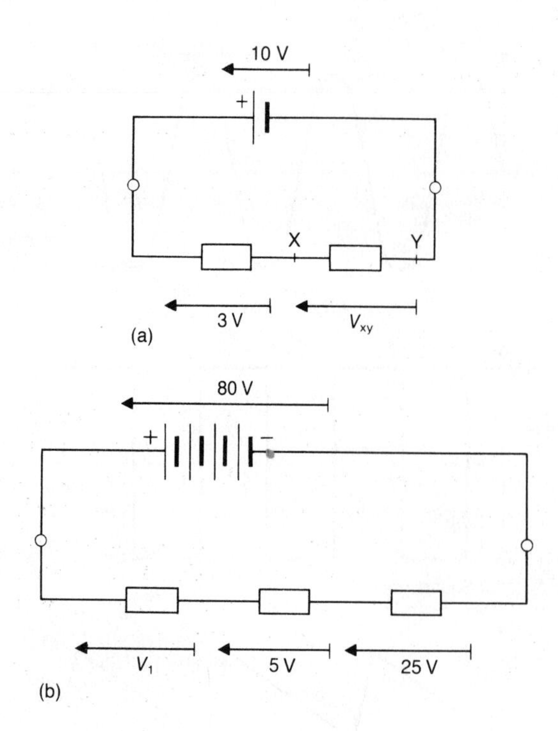

Figure 10.10 Circuits for problem 4

4 Using the important result, often known as Kirchhoff's voltage law:

> Applied source voltage = sum of voltages across the individual components making up the circuit

determine the unknown voltages marked in on the circuits of Figure 10.10.

11 Resistance and series and parallel resistive circuits

General learning objectives: to solve problems related to current, potential difference and resistance for simple resistive circuits in series and in parallel.

11.1 Resistance and resistors

The British Standard (BS) definition of **resistance** is:

1 Resistance is that property of a substance which restricts the flow of electricity through it associated with the conversion of electrical energy to heat.
2 The magnitude of this restriction.

The term **resistor** is given to components which have the property of resistance in that they limit the flow of electrical current and dissipate electrical energy as heat. It is important to understand resistance as the property of restricting current flow together with the conversion of electrical energy to heat, since there are other electrical components which can restrict current flow without necessarily converting electrical energy to heat, e.g. inductors and capacitors can restrict alternating current flow without converting any electrical energy to heat.

The quantitative value of resistance of a resistor component is defined as

$$R = \frac{V}{I}$$

where R is the letter used to denote resistance, and
V = voltage across the terminals of the component, units volts (V)
I = current flowing through the component, units amperes (A)

The SI units of resistance, volts per ampere, are known as ohms, denoted by the Greek letter omega, Ω:

$$1\,\text{ohm} = \frac{1\,\text{volt}}{1\,\text{ampere}}, \quad \Omega = \text{V/A}$$

Some examples illustrating the construction of practical resistors are shown in Figure 11.1. The carbon composition type shown in (a) is used extensively in electronic circuits. It consists basically of a carbon rod with conducting wire leads firmly attached to its ends. A wire-wound resistor is shown in (b); wire-wound resistors consist of resistive wire, such as manganin or constantan, wound on an insulating former and usually covered with a layer of vitreous enamel to

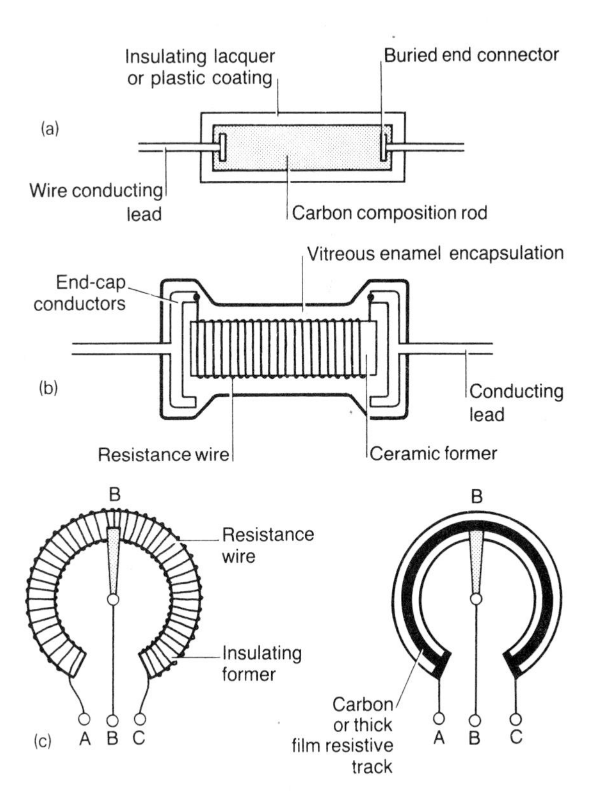

Figure 11.1 Examples of some practical resistors
(a) Carbon compostion type resistor
(b) Wire-wound type resistor
(c) Wire-wound and resistive track variable resistors

protect and also to help conduct heat away from the component.

Two examples of variable resistor are shown in (c). They consist essentially of either bare resistance wire wound on an insulator former or a carbon or thick film resistive track between two fixed terminals, A and C. B is a movable spring contact whose position can be varied manually enabling the resistance between A and B to be altered from zero and the full value of the complete resistor coil of wire or the resistive track as B is moved from A to C.

Example

Determine the resistance of the component R in each of the circuits shown in Figure 11.2.

Solution

Since the voltage V across the resistor and current I are given in each case we can apply the definition $R = V/I$ in each case.

(a) $V = 50\,\text{V}, I = 2\,\text{A}$

$$\text{so } R = \frac{V}{I} = \frac{50}{2} = 25\,\Omega \quad \textit{Ans}$$

(b) $V = 20\,\text{V}, I = 5\,\text{mA} = 0.005\,\text{A}$

$$\text{so } R = \frac{20}{0.005} = \frac{20\,000}{5}$$

$$= 4000\,\Omega \text{ or } 4\,\text{k}\Omega \quad \textit{Ans}$$

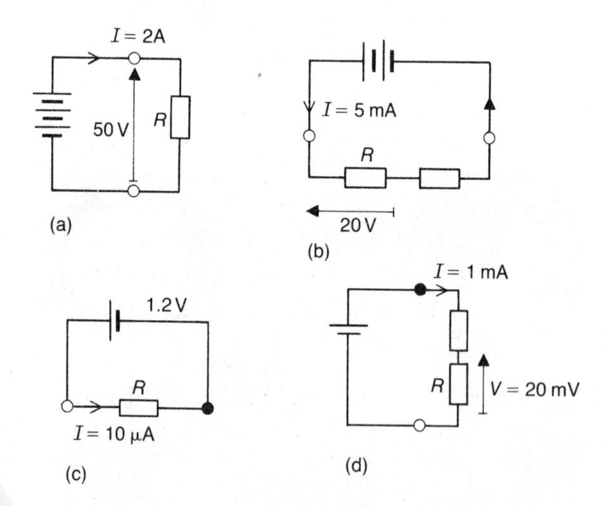

Figure 11.2

(c) $V = 1.2\,\text{V}, I = 10\,\mu\text{A} = 10 \times 10^{-6}\,\text{A}$

$$\text{so } R = \frac{1.2}{10 \times 10^{-6}} = \frac{1.2}{10^{-5}}$$

$$= 1.2 \times 10^{5}\,\Omega \text{ or } 120\,\text{k}\Omega \quad \textit{Ans}$$

(d) $V = 20\,\text{mV} = 20 \times 10^{-3}\,\text{V}$
$I = 1\,\text{mA} = 1 \times 10^{-3}\,\text{A}$

$$\text{so } R = \frac{20 \times 10^{-3}}{1 \times 10^{-3}} = 20\,\Omega \quad \textit{Ans}$$

11.2 Ohm's law

The relationship between the voltage or potential difference across a conductor and the current flowing through it was first experimentally investigated by Georg Simon Ohm (1787–1854). Ohm observed that at a constant temperature the current flowing through a conductor was directly proportional to the voltage across it. This relationship, which is perhaps the most famous and most widely used 'law' in electrical circuits, carries his name and is known as Ohm's Law.

Ohm's Law may be stated as:

The current I flowing in a conductor is directly proportional to the voltage difference V between its ends, provided physical conditions, such as the temperature, remain constant.

The law may be expressed by the formula

$$I = \frac{V}{R} \text{ or } V = RI$$

where the constant of proportionality R is the resistance of the conductor.

It should be noted that Ohm's law is an experimental law and although it is closely obeyed by many conducting materials it is not universally applicable. Most metallic conductors, some non-metallic conductors and some salt and other electrolytic solutions closely satisfy Ohm's law. However, some good conductors of electric current and semiconductor materials (used to make diodes, transistors and integrated circuits) do not. Conductors which obey Ohm's law are termed linear, i.e. $I = V/R$ or $V = RI$ is a linear (straight line) equation. Conductors which do not obey Ohm's law are termed non-linear.

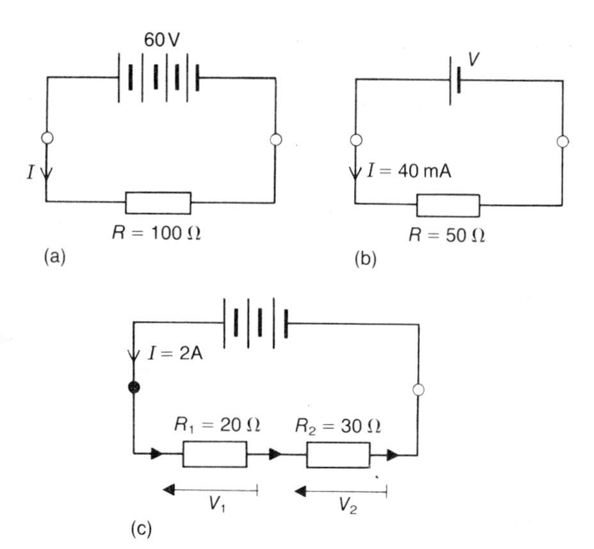

Figure 11.3 For example 1

Examples

1 Using Ohm's law determine

(a) the current I in the circuit of Figure 11.3(a);

(b) the value of the source e.m.f. V in Figure 11.3(b);

(c) the values of the potential differences, V_1 and V_2 across the resistors $R_1 = 20\,\Omega$, $R_2 = 30\,\Omega$, in the circuit of Figure 11.3(c).

Solution

(a) The voltage developed across the $100\,\Omega$ resistor is equal to the source voltage. Hence using Ohm's law with $V = 60\,\text{V}$ and $R = 100\,\Omega$,

$$\text{current } I = \frac{V}{R} = \frac{60}{100} = 0.6\,\text{A} \quad \textit{Ans}$$

(b) Source e.m.f., $V = RI$

where $R = 50\,\Omega$ and $I = 40\,\text{mA} = 0.040\,\text{A}$,

hence $V = 50 \times 0.040 = 2\,\text{V}$ *Ans*

(c) Voltage across R_1,

$$V_1 = R_1 I = 20 \times 2 = 40\,\text{V} \quad \textit{Ans}$$

Voltage across R_2,

$$V_2 = R_2 I = 30 \times 2 = 60\,\text{V} \quad \textit{Ans}$$

Note: the current I flows through both R_1 and R_2.

2 The following results were obtained for the voltage V across and the current I flowing through two components.

Component A

V(V)	−5	−4	−3	−2	−1	0	+1	2	3	4	5
I(mA)	−100	−80	−60	−40	−20	0	20	40	60	80	100

Component B

V(V)	−1	−0.5	0	+0.5	0.6	0.7	0.8	0.9	1	1.1
I(mA)	0	5	0	0.08	1	3	7.5	16	26	40

Draw the graphs of the relationship of current versus voltage for components A and B and hence state which component is a linear and which is a non-linear resistance element.

Determine (a) the current through component A when the voltage across A is equal to 3.5 V, (b) the voltage across component B when the current flowing through it is 20 mA.

Solution

The graph of I versus V for component A is drawn in Figure 11.4(a) and the graph for component B is drawn in Figure 11.4(b). The relationship between I and V for component A is a straight line, hence component A is a linear element. Component B is a non-linear element since Figure 11.4(b) shows that the plot of I versus V is not a straight line.

(a) We may find the current through A when $V = 3.5\,\text{V}$ by using the I v. V graph. First locate 3.5 V on the voltage axis and draw a vertical line as shown in Figure 11.4(a). This line cuts the I–V straight line of component A at the point Q (also shown). Draw a horizontal line through Q and read off value of current where this line cuts I axis. From Figure 11.4(a), this gives

$I = 70\,\text{mA}$, the current through A, when $V = 3.5\,\text{V}$ *Ans*

(b) On the I v. V curve of Figure 11.4(b) draw horizontal line through $I = 20\,\text{mA}$. This line cuts the curve at P. Draw vertical line through P and read off value of V where it cuts the voltage axis. This gives (as shown in Figure 11.4(b))

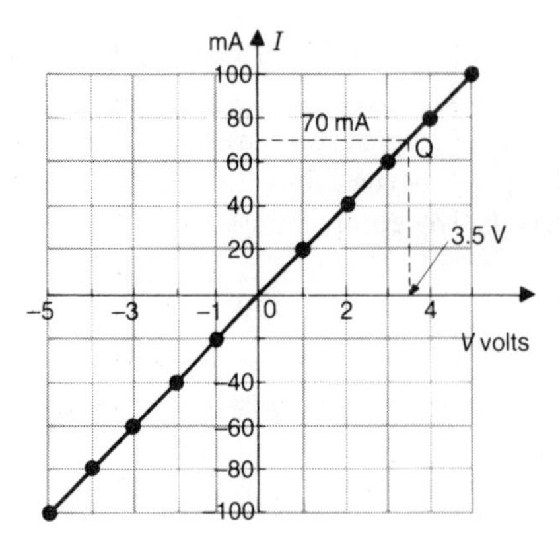

(a) I v V graph for component A

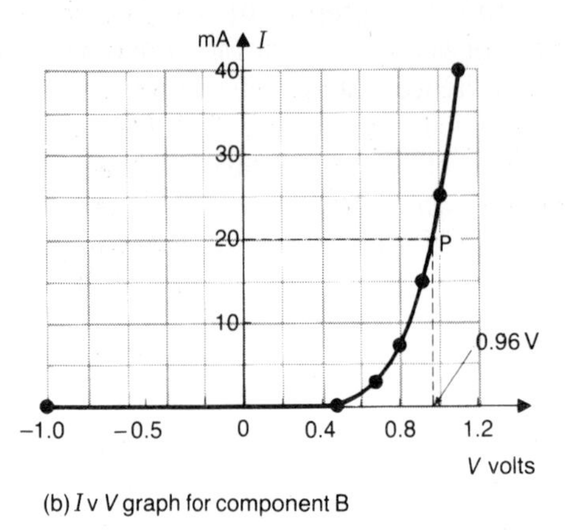

(b) I v V graph for component B

Figure 11.4 For example 2

$V = 0.96\,\text{V}$ (approximately), when $I = 20\,\text{mA}$ *Ans*

11.3 The measurement of current and voltage: ammeters and voltmeters

An instrument that measures electrical current is known as an ammeter. An ammeter has two terminals, one normally marked with a plus (+) or coloured red to indicate the positive polarity terminal, and one marked with with a minus (−) and often coloured black to indicate the negative polarity terminal. The ammeter should always be connected so that dc current enters at the positive terminal and leaves at the negative terminal. The ammeter has a scale suitably graduated in amperes or multiple or sub-multiple units, such as milliamperes (mA) or for very sensitive current measurement, microamperes (μA).

An instrument that measures voltage or potential difference is known as a voltmeter and is similar in appearance to an ammeter, except, of course that its scale is calibrated in volts or multiples or sub-multiples of volts, such as kilovolts (kV) for very high-voltage work or millivolts (mV) for low-voltage measurements. A voltmeter has two terminals, the positive and negative polarity terminals and in measurements the positive (+) terminal should always be connected to the higher voltage point.

Sketches of some common forms of ammeter, a multi-range voltmeter, a digital multimeter and the circuit symbols for an ammeter and voltmeter are drawn in Figure 11.5.

To measure the current in a circuit, we must break the circuit at a suitable point and connect the + terminal of the ammeter to the more positive side (higher voltage side) of the circuit and the − terminal to the other side. Examples of the connection of an ammeter to measure current in the various components making up the circuit of Figure 11.6(a) are shown in (b), (c) and (d). Note that the ammeter is always connected in series with the component.

To measure the voltage across the circu[illegible] component, we connect the voltmeter across the component (i.e. in parallel with the component) with the + terminal at the higher voltage side and the − terminal connected to the lower voltage side. Examples of the connection of a voltmeter to measure the voltage across the various components of Figure 11.7 are shown in (b) and (c).

Ideally, an ammeter should have zero resistance so that when it is connected in series in [illegible] circuit, zero voltage is dropped across its terminals. Its inclusion should not affect a circuit in any way. On the other hand, an ideal voltmeter should have infinite resistance so that when [illegible] voltmeter is connected across a circuit element, [illegible] will draw no current and therefore not disturb the conditions in the circuit.

Modern electronic voltmeters closely approa[illegible] the ideal instrument in that they present [illegible] resistance typically in excess of 10 MΩ. Howeve[illegible] voltmeters incorporating a moving-coil instr[illegible]

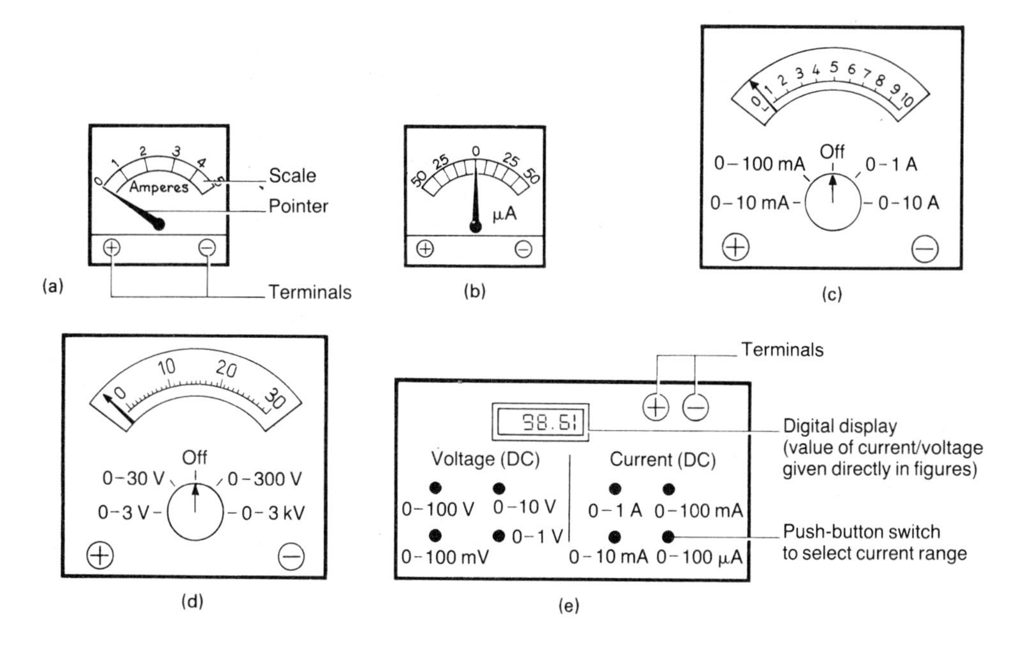

Figure 11.5 Ammeters and voltmeters
(a) DC ammeter
(DC = direct current)
(b) Centre-zero microammeter
(c) Multi-range ammeter
(d) Multi-range voltmeter
(e) Digital multimeter which can measure both voltage and give number (digital) display
(f) Circuit symbols

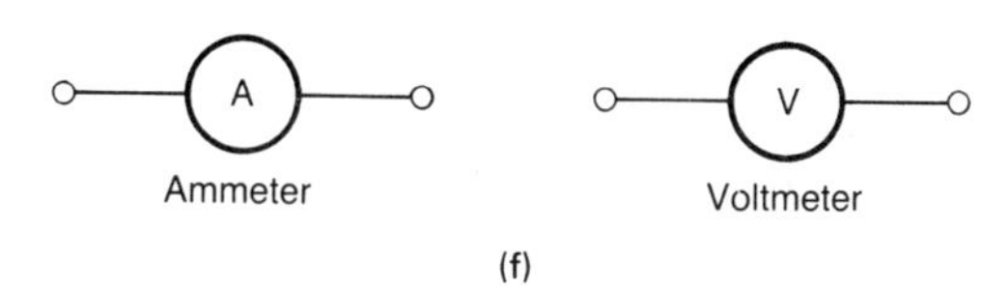

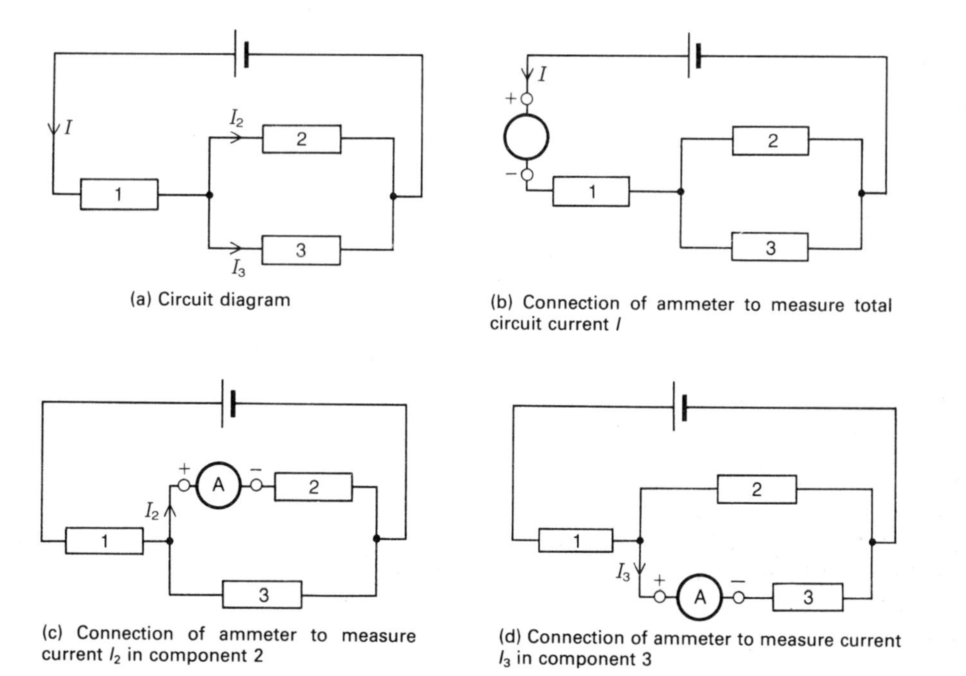

Figure 11.6 Connecting an ammeter to measure current

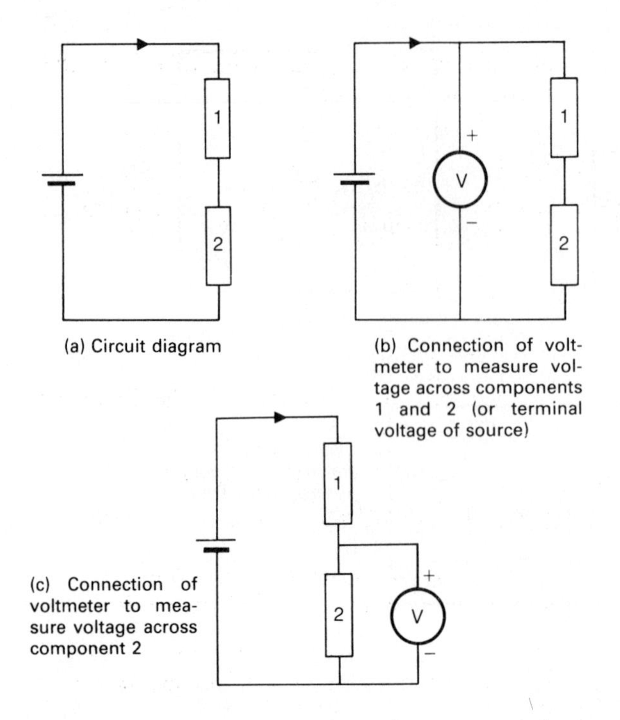

Figure 11.7 Connecting a voltmeter to measure voltage

ment present a high, but finite, resistance, and thus, in certain voltage measurements, the instrument may draw a current comparable with that flowing in the circuit. Under these circumstances, we must be aware that our measurements may be in error. Practical ammeters have a small resistance, and thus a small voltage, typically in the range of 100 mV to 500 mV, is dropped across their terminals when measuring current in a circuit. In most applications, this will lead to negligible error, but care must be exercised where such small voltage drops are comparable with those being measured.

11.4 The fundamental circuit laws: Kirchhoff's current and voltage laws

The current and voltage laws, known collectively as Kirchhoff's laws and named after Gustav Kirchhoff (1824–87), together with Ohm's law, form the basis of circuit analysis.

The **current law**:

The algebraic sum of the electric currents which meet at any point in a circuit is zero.

This law expresses the physical requirement that charge must be conserved at all points in a circuit so that the sum of all currents entering at any point equals the sum of the currents leaving that point. For example, in the circuit of Figure 11.8,

current entering the junction point, P is I;
currents leaving P are I_1 and I_2;
so $I = I_1 + I_2$

or, taking into account the directions of the currents and assuming currents entering a point are regarded as positive and so currents marked as leaving the point are negative,

algebraic sum $= I - I_1 - I_2 = 0$

The **voltage law**:

The algebraic sum of the electromotive forces in any closed circuit (loop) is equal to the sum of the potential differences across the elements making up the circuit (loop).

(Note: a *loop* is a combination of circuit elements forming a closed path in a circuit. In Figure 11.8 the battery and elements R_A and R_B form a *loop*; the battery and elements R_A and R_C also form a *loop*.)

This law essentially expresses the law of conservation of energy in a circuit. For resistive element circuits the law may simply expressed as:

Sum of applied e.m.f.s = sum of products of the resistances of the elements and the current flowing through them

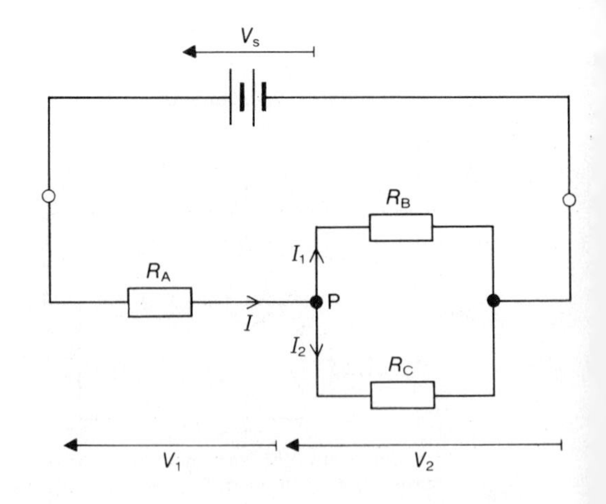

Figure 11.8

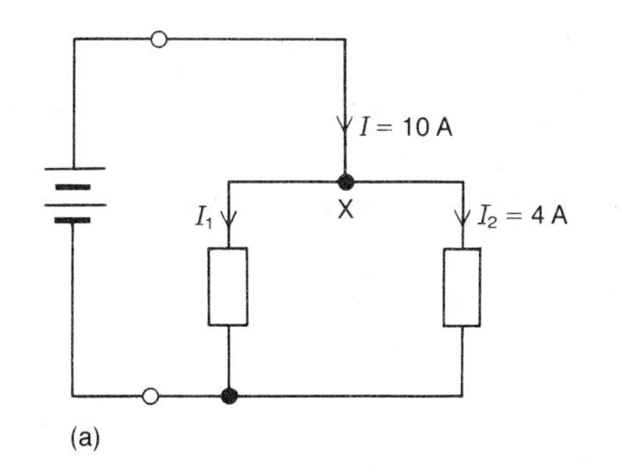

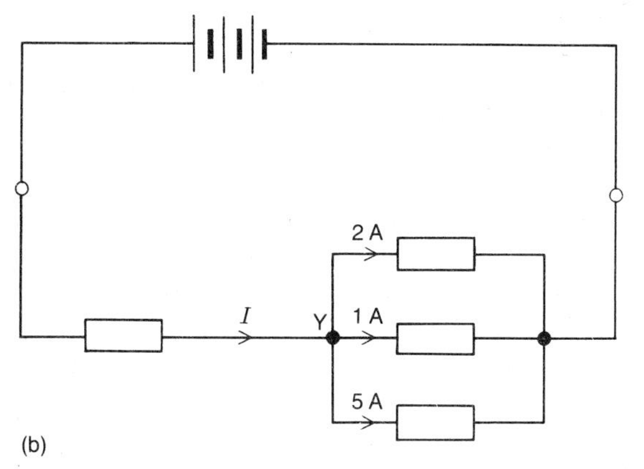

Figure 11.9 Circuits for example 1

For example, for the circuit of Figure 11.8,

applied e.m.f $= V_s$
voltage or potential difference across $R_A = V_1$
voltage across R_B and also $R_C = V_2$

so applying the voltage law,

$V_s = V_1 + V_2$

and incorporating also Ohm's law,

i.e. $V_1 = R_A I$, $V_2 = R_B I_1$
and also $V_2 = R_C I_2$
we have $V_S = R_A I + R_B I_1$
(for $V_S - R_A - R_B$ loop)
$V_S = R_A I + R_C I_2$
(for $V_S - R_A - R_C$ loop)

Examples

1 Using the current law:

sum of currents flowing in = sum of currents flowing out

at any junction point in a circuit, determine the unknown currents marked in on the circuits of Figure 11.9.

Solution

(a) At the junction point X in the circuit of Figure 11.9(a),

$$I = I_1 + I_2$$
$$10 = I_1 + 4$$
$$\text{so } I_1 = 10 - 4 = 6\,\text{A} \quad \textit{Ans}$$

(b) At the junction point Y in the circuit of Figure 11.9(b)

$$I = 2 + 1 + 5 = 8\,\text{A} \quad \textit{Ans}$$

2 Using the voltage law:

algebraic sum of applied e.m.f.s = sum of voltages across individual components

in any closed path (loop) in a circuit, determine the unknown voltages shown in the circuits of Figure 11.10.

Solution

(a) In the circuit of Figure 11.10(a),

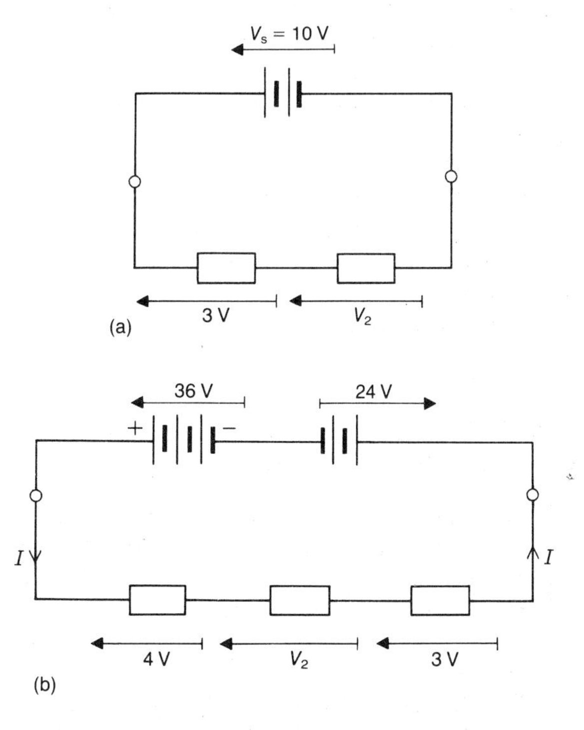

Figure 11.10 Circuits for example 2

applied e.m.f. V_s = sum of voltage across the two components

i.e. $10 = 3 + V_2$

so $V_2 = 10 - 3 = 7\,V$ *Ans*

(b) In the circuit of Figure 11.10(b) we have two sources acting in opposition to each other,

algebraic sum of e.m.f.s = $36 - 24 = 12\,V$

sum of voltages across elements = $4 + V_2 + 3 = 7 + V_2$

so $12 = 7 + V_2$

$V_2 = 12 - 7 = 5\,V$ *Ans*

11.5 Series and parallel combinations of resistors: equivalent resistance

It is often very useful in circuit calculations to find the equivalent resistance or resultant resistance a

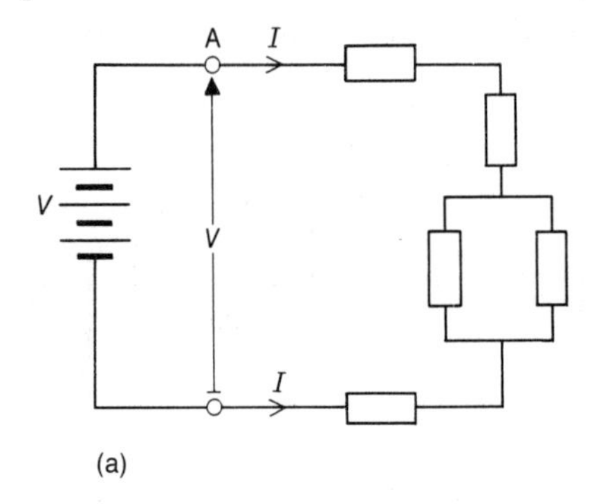

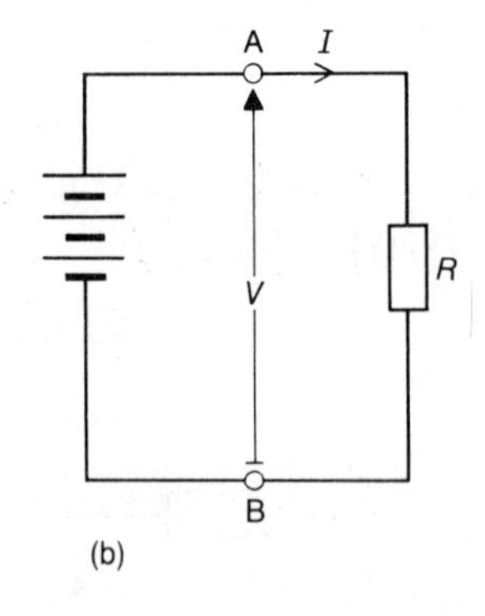

Figure 11.11 Equivalent resistance R *replaces the total effect of the resistor circuit and so far as the source is concerned is identical to original circuit*

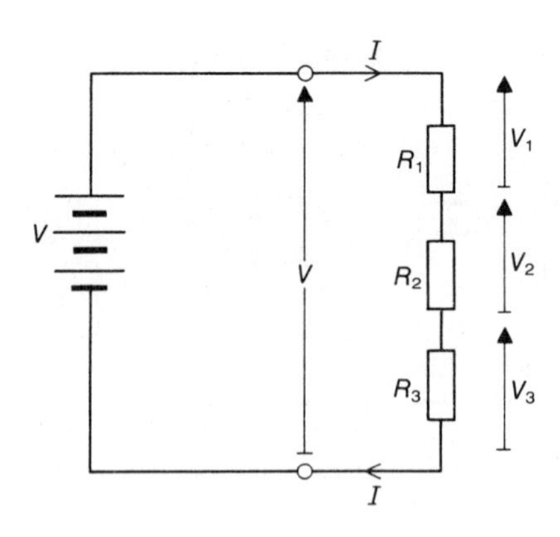

Figure 11.12 Series circuit: equivalent resistance R = R$_1$ + R$_2$ + R$_3$

circuit presents to the terminals of the voltag source or electrical generator. The equivalen resistance of a circuit consisting of resistors is th value of the single resistor which when it replace the resistor circuit causes the identical current t flow from the source. For example, in Figu 11.11 the equivalent resistance of the circuit in (a looking at terminals A–B is,

$$R = \frac{V}{I} \text{ ohms}$$

where V = voltage across terminals A–B
I = current flowing in and out of thes terminals

11.5.1 Equivalent resistance of a series combination of resistors

For a series circuit,

equivalent resistance R = sum of the individual resistanc

For example, for the circuit of Figure 11.12,

$$R = R_1 + R_2 + R_3$$

The result may easily be proved using Ohm's lav and the properties of a series circuit:

(a) the current is the same in all parts of a series circuit;
(b) the voltage law: applied voltage = sum o individual voltages.

So for the case of Figure 11.12,

$$V = V_1 + V_2 + V_3$$
$$= R_1 I + R_2 I + R_3 I$$
$$= (R_1 + R_2 + R_3)I$$

so $$R = \frac{V}{I} = R_1 + R_2 + R_3$$

e.g. if $R_1 = 22\,\Omega$, $R_2 = 47\,\Omega$ and $R_3 = 68\,\Omega$
then $R = 22 + 47 + 68 = 137\,\Omega$

11.5.2 Equivalent resistance of a parallel combination of resistors

For a parallel circuit,

$$\frac{1}{R} = \frac{1}{R_1} + \frac{1}{R_2} + \dots$$

The equivalent resistance R is found by adding the **reciprocals** of the individual resistance values.

For example, in the circuit of Figure 11.13(a), where $R_1 = 20\,\Omega$, $R_2 = 30\,\Omega$, we have

$$\frac{1}{R} = \frac{1}{R_1} + \frac{1}{R_2}$$
$$= \frac{1}{20} + \frac{1}{30} = \frac{3+2}{60} = \frac{5}{60}$$

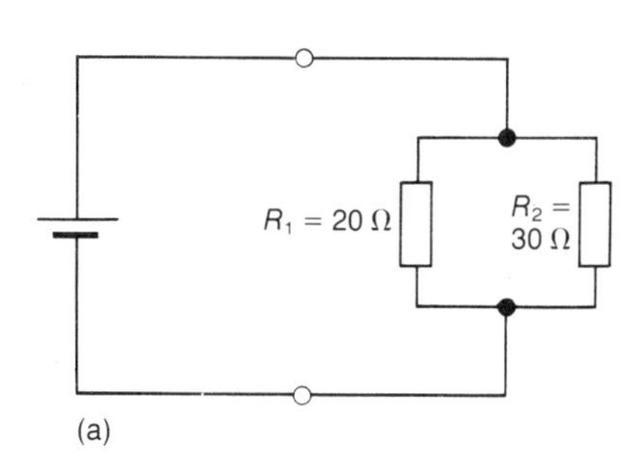

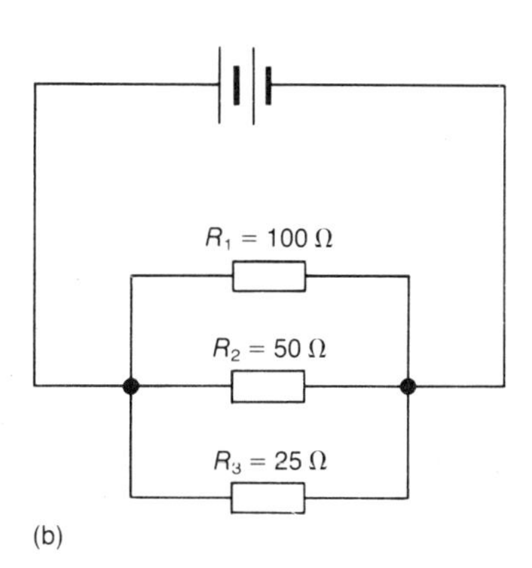

Figure 11.13 Parallel circuits

so $$R = \frac{60}{5} = 12\,\Omega$$

In the circuit of Figure 11.13(b),

$$\frac{1}{R} = \frac{1}{R_1} + \frac{1}{R_2} + \frac{1}{R_3}$$
$$= \frac{1}{100} + \frac{1}{50} + \frac{1}{25}$$
$$= \frac{1+2+4}{100} = \frac{7}{100}$$

so $$R = \frac{100}{7} = 14.3\,\Omega$$

The equivalent resistance formula for parallel circuits follows from the current law, i.e. the total current entering the parallel combination equals the sum of the currents flowing in the individual parallel resistors; and the fact that the voltage across the parallel connected resistors is the same. Thus, for the parallel circuit of Figure 11.4,

$$I = I_1 + I_2 + I_3 + I_4$$
$$= \frac{V}{R_1} + \frac{V}{R_2} + \frac{V}{R_3} + \frac{V}{R_4}$$
$$= V\left(\frac{1}{R_1} + \frac{1}{R_2} + \frac{1}{R_3} + \frac{1}{R_4}\right)$$

so $$\frac{I}{V} = \frac{1}{R} = \frac{1}{R_1} + \frac{1}{R_2} + \frac{1}{R_3} + \frac{1}{R_4}$$

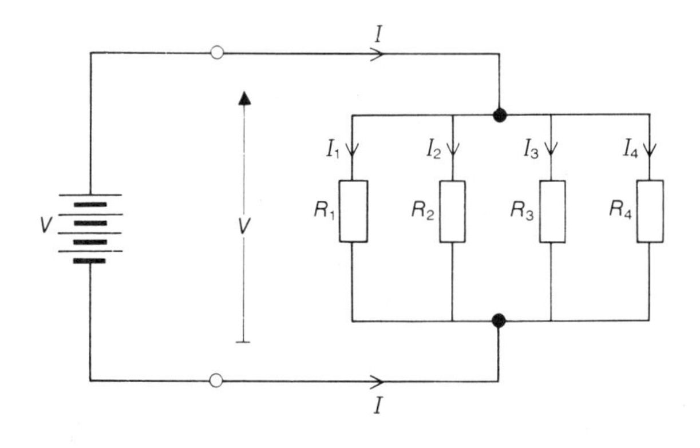

Figure 11.14 Parallel resistor circuit: $\frac{1}{R} = \frac{1}{R_1} + \frac{1}{R_2} + \frac{1}{R_3} + \frac{1}{R_4}$

11.6 Solution of problems for series and parallel resistive circuits

We can solve resistive circuit problems by applying the voltage and current laws, together with Ohm's law. The rules for combining series and parallel combinations of resistors are also extremely useful and often save a considerable amount of time in effecting a solution.

Examples

1 Determine the current I and the voltages across the individual resistors in the series circuit of Figure 11.15.

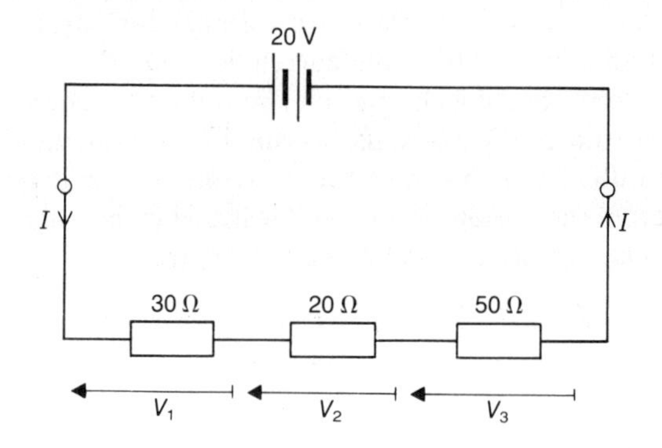

Figure 11.15 Circuit for example 1

Solution

The total resistance of the circuit,

$$R = 30 + 20 + 50 = 100\,\Omega$$

hence the current flowing,

$$I = \frac{V}{R} = \frac{20}{100} = 0.2\,\text{A} \quad \textit{Ans}$$

as the applied e.m.f. $V = 20\,\text{V}$.

The voltages across the individual resistors are:

$$V_1 = 30I = 30 \times 0.2 = 6\,\text{V}$$
$$V_2 = 20I = 20 \times 0.2 = 4\,\text{V}$$
$$V_3 = 50I = 50 \times 0.2 = 10\,\text{V} \quad \textit{Ans}$$

2 The current measured by the ammeter in the series circuit of Figure 11.16 is $I = 2\,\text{A}$ and the voltage measured across resistor R_1 is $V_1 = 16\,\text{V}$. Calculate:

(a) the total resistance of the circuit;

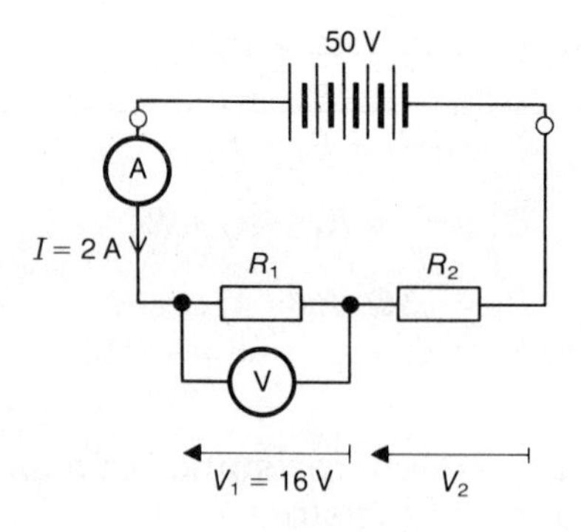

Figure 11.16 Circuit for example 2

(b) the values of R_1 and R_2;

(c) the voltage V_2.

Solution

(a) The total resistance of the circuit, R

$$R = \frac{\text{applied e.m.f.}}{\text{current}} = \frac{50}{2} = 25\,\Omega \quad \textit{An}$$

(b) The value of R_1,

$$R_1 = \frac{V_1}{I} = \frac{16}{2} = 8\,\Omega \quad \textit{Ans}$$

and since $R = R_1 + R_2$

$$R_2 = R - R_1 = 25 - 8 =$$

A

(c) The voltage V_2 across R_2,

$$V_2 = R_2 I = 17 \times 2 = 34\,\text{V} \quad \textit{Ans}$$

which checks with,

$$V_2 = \text{e.m.f.} - V_1 = 50 - 16 = 34\,\text{V}$$

3 Calculate the currents I_1, I_2 and I in th parallel resistor circuit of Figure 11.17.

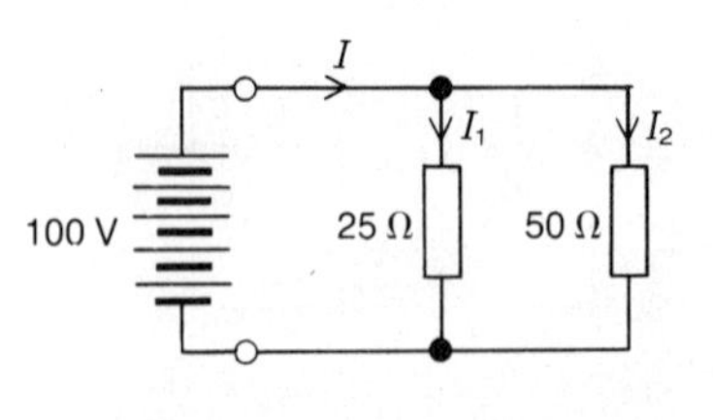

Figure 11.17 Circuit for example 3

Solution

The voltage V across the paralleled resistors is equal to the applied e.m.f. of 100 V, so

$$I_1 = \frac{V}{25} = \frac{100}{25} = 4\text{ A} \quad Ans$$

$$I_2 = \frac{V}{50} = \frac{100}{50} = 2\text{ A} \quad Ans$$

and using the current law,

$$I = I_1 + I_2 = 4 + 2 = 6\text{ A} \quad Ans$$

4 For the circuit of Figure 11.18, calculate
(a) the total resistance and supply current I;
(b) the voltage across the 10 Ω resistor;
(c) the current I_1 in the 60 Ω resistor.

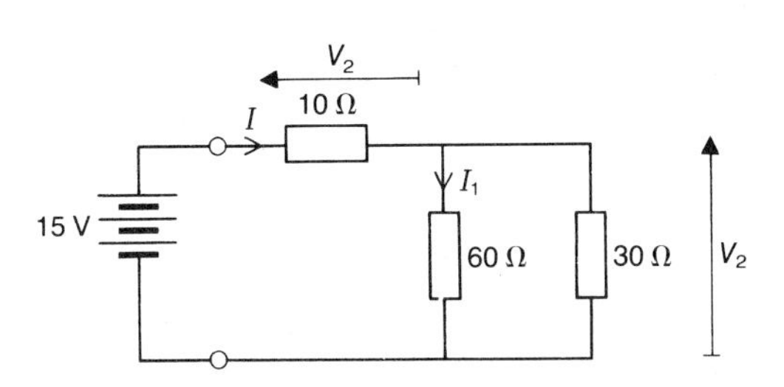

Figure 11.18 Circuit for example 4

Solution

(a) The total resistance of the circuit consists of 10 Ω plus the equivalent resistance, R_2 say, of 60 Ω in parallel with 30 Ω:

$$\frac{1}{R_2} = \frac{1}{60} + \frac{1}{30} = \frac{1+2}{60} = \frac{3}{60}$$

$$\text{so } R_2 = \frac{60}{3} = 20\,\Omega$$

and the total resistance,

$$R = 10\,\Omega \text{ in series with } 20\,\Omega = 30\,\Omega$$

Ans

The supply current,

$$I = \frac{\text{e.m.f.}}{R} = \frac{15}{30} = 0.5\text{ A} \quad Ans$$

(b) The voltage across the 10 Ω resistor,

$$V_1 = 10 \times I = 10 \times 0.5 = 5\text{ V} \quad Ans$$

(c) The voltage across the parallel resistors,

$$V_2 = \text{e.m.f.} - V_1 = 15 - 5 = 10\text{ V}$$

so the current in the 60 Ω resistor,

$$I_1 = \frac{V_2}{60} = \frac{10}{60} = \frac{1}{6}$$

or 0.167 A *Ans*

5 Calculate the currents I and I_1, I_2 and I_3 in the circuit of Figure 11.19.

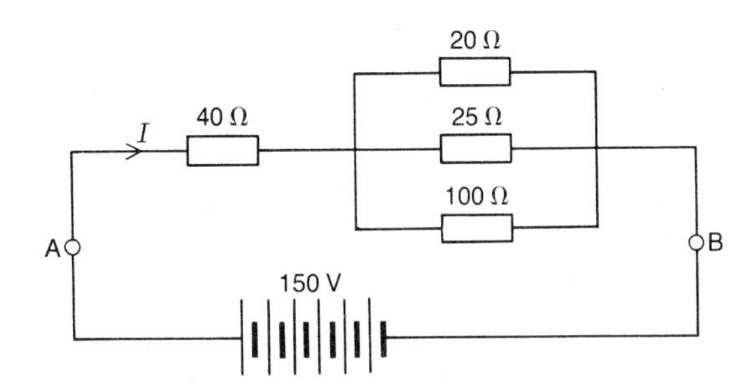

Figure 11.19 Circuit for example 5

Solution

First work out total resistance of circuit,

$$R = 40\,\Omega + (20\,\Omega, 25\,\Omega, 100\,\Omega \text{ in parallel})$$

Suppose the equivalent parallel resistance is R_2, then

$$\frac{1}{R_2} = \frac{1}{20} + \frac{1}{25} + \frac{1}{100} = \frac{5+4+1}{100} = \frac{10}{100}$$

$$\text{so} \quad R_2 = \frac{100}{10} = 10\,\Omega$$

and therefore, $R = 40\,\Omega + 10\,\Omega = 50\,\Omega$

Hence the supply current,

$$I = \frac{\text{applied voltage across AB}}{R} = \frac{150}{50} = 3\text{ A}$$

Ans

The voltage drop across the 40 Ω resistor, V_1 say, is

$$V_1 = 40 \times I = 40 \times 3 = 120\text{ V}$$

so the voltage across the three parallel resistors, V_2 say, is

$$V_2 = 150 - 120 = 30\,\text{V}$$

or alternatively,

$$V_2 = R_2 I = 10 \times 3 = 30\,\text{V}$$

Hence current in $20\,\Omega$ resistor $= \dfrac{V_2}{20} = \dfrac{30}{20} = 1.5\,\text{A}$

current in $25\,\Omega$ resistor $= \dfrac{V_2}{25} = \dfrac{30}{25} = 1.2\,\text{A}$

current in $100\,\Omega$ resistor $= \dfrac{V_2}{100} = \dfrac{30}{100} = 0.3\,\text{A}$ *Ans*

11.7 The effect of temperature on resistance

The resistance of most conductors of electricity varies with temperature. The resistance of most metals and metal alloys increases with a rise in temperature. However, certain conductors have a resistance which actually decreases with temperature rise. Carbon, most semiconductor materials and electrolytic solutions have a resistance which decreases as temperature increases.

Over a limited temperature range, the resistance of a conductor is given quite accurately by the formula,

$$R_T = R_0(1 + \alpha T)$$

where R_T = resistance of the conductor at T°C
R_0 = resistance of the conductor at 0°C
α = a constant, known as *the temperature coefficient of resistance*

Making α the subject of the above formula, we have

$$R_T = R_0 + R_0\alpha T$$

$$R_T - R_0 = R_0\alpha T$$

so
$$\alpha = \frac{R_T - R_0}{R_0 T}$$

i.e.
$$\alpha = \frac{\text{resistance change}}{\text{original resistance} \times \text{temperature change}}$$

Values of α for some commonly used conductors are given in Table 11.1.

Note that the coefficients for manganin, constantan and nickel–chromium are very much smaller than for the non-alloy metals. Manganin and constantan are often used for producing standard wire-wound resistors since their resistance change with temperature can be made very small. The variation of resistance with temperature of platinum is used in the plantinum resistance thermometer for the precision measurement of temperature over the range −150°C to above 1000°C.

Examples

1 The resistance of a long length of copper wire is measured at 0°C and also at 40°C and the resistance increase found to be $8.6\,\Omega$. If the resistance at 0°C is $50\,\Omega$, determine the temperature coefficient of resistance for copper over the range 0 to 40°C.

Solution

The temperature coefficient of resistance,

Table 11.1 Temperature coefficient of resistance for some commonly used conductor materials

Material	*Temperature coefficient of resistance* (°C^{-1})
copper	0.004
iron	0.006
aluminium	0.004
platinum	0.004
silver	0.004
gold	0.003
steel	varies considerably
carbon	−0.0005
manganin (84% Cu, 4% Ni, 12% Mn)	±0.00001
constantan (60% Cu, 40% Ni)	±0.00004
nickel–chromium (80% Ni, 20% Cr)	±0.0001

Temperature range 0–100°C approximately

$$\alpha = \frac{\text{resistance change}}{\text{original resistance} \times \text{temperature change}}$$

$$= \frac{8.6}{50 \times 40} = 0.0043\,°\text{C}^{-1} \quad \textit{Ans}$$

2 Calculate the resistance of a carbon resistor at 45 °C if its resistance at 0°C is 1 kΩ and its temperature coefficient of resistance, $\alpha = -0.0005\,°\text{C}^{-1}$

Solution

The resistance at T = 45°C is given by,

$$R_T = R_0(1 + \alpha T)$$

where $R_0 = 1\,\text{k}\Omega = 1000\,\Omega$

and $\alpha = -0.0005\,°\text{C}^{-1}$

so $R_T = 1000[1 - (0.0005 \times 45)]$

$= 977.5\,\Omega$ *Ans*

3 A temperature-sensitive resistor element is used to monitor temperature. Its resistance at 0°C is 100 Ω and its temperature coefficient of resistance, $\alpha = 0.005\,°\text{C}^{-1}$. Calculate the temperature of the element when its resistance is

(a) 94.5 Ω; (b) 110.5 Ω.

Solution

Solving the general formula,

$$R_T = R_0(1 + \alpha T)$$

for temperature T, we have

$$R_T = R_0 + \alpha R_0 T$$
$$R_T - R_0 = \alpha R_0 T$$
$$\text{so } T = \frac{R_T - R_0}{\alpha R_0}$$

(a) When $R_T = 94.5$ with $R_0 = 100$, we obtain

$$T = \frac{94.5 - 100}{0.005 \times 100}$$

$$= \frac{-5.5}{0.5} = -11°\text{C} \quad \textit{Ans}$$

(b) When $R_T = 110.5$,

$$T = \frac{110.5 - 100}{0.005 \times 100} = +21°\text{C} \quad \textit{Ans}$$

Test 11

This test may be used as a basic self-assessment test to check your understanding of Chapter 11 on **Resistance and series and parallel resistive circuits**, and its learning objectives. Enter all answers in the answer block.

Qu. 1 Enter a tick (√) in the answer block if you consider the statement is correct; enter a cross (×) if you consider the statement is in any way incorrect.

(a) Ohm's law states that at constant temperature the current flowing through a conductor is directly proportional to the voltage across its terminals.

(b) All conductors of electricity obey Ohm's law.

(c) The current I leaving the junction point in Figure 11.20 is 18 A.

(d) The voltage V_1 across the 40 Ω resistor in the circuit of Figure 11.21 is 120 V and the current I = 3 A.

Answer block:

Question	1				2		3		
	(a)	(b)	(c)	(d)	(a)	(b)	(a)	(b)	(c)
Answer									

Question	4			5			6	
	(a)	(b)	(c)	(a)	(b)	(c)	(a)	(b)
Answer								

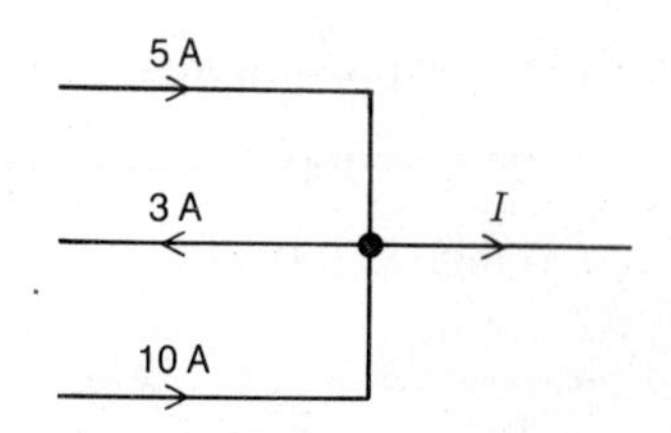

Figure 11.20 For Qu 1(c)

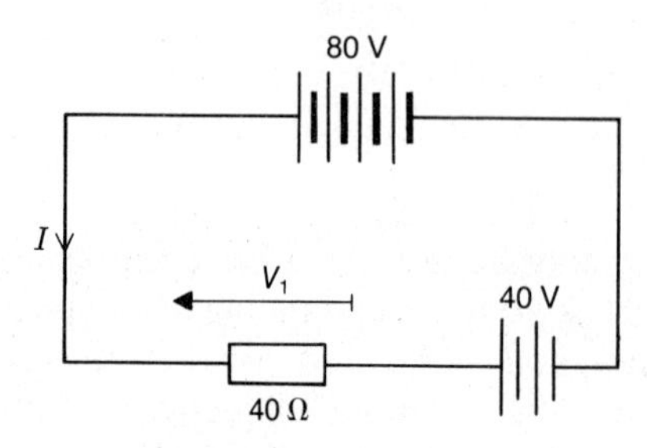

Figure 11.21 For Qu 1(d)

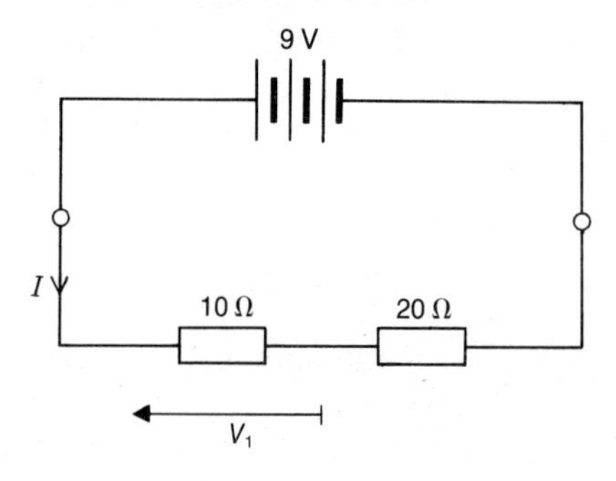

Figure 11.22 For Qu 2

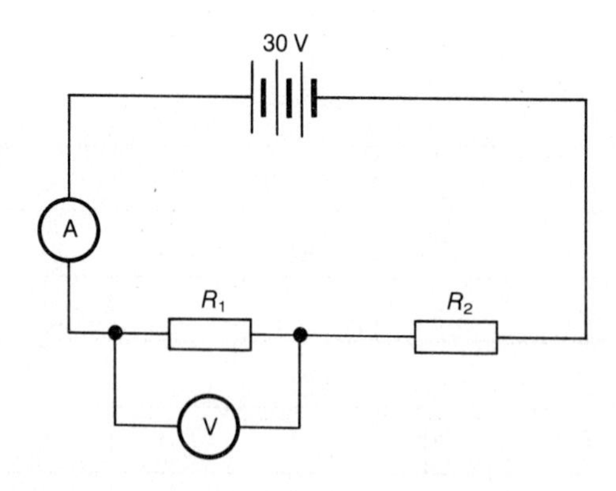

Figure 11.23 For Qu 3

Qu. 2 Calculate, for the circuit shown in Figure 11.22:
(a) the current I;
(b) the voltage V_1.

Qu. 3 In the circuit shown in Figure 11.23 the current measured by the ammeter is 0.5 A and the voltage measured by the voltmeter is 10 V. Determine:
(a) the value of resistor R_1;
(b) the voltage across resistor R_2;
(c) the value of resistor R_2.

Qu. 4 Calculate, for the circuit shown in Figure 11.24, the currents
(a) I_1; (b) I_2; (c) I.

Qu. 5 Calculate the equivalent resistance between terminals A–B for the resistor combinations of Figure 11.25(a), (b) and (c).

Qu. 6 The resistance of a coil of wire measured at a temperature of 0°C is 100 Ω and when remeasured at a temperature of 20°C is 104 Ω. Calculate:
(a) the temperature coefficient of resistance of the wire material;
(b) the resistance of the coil at 50°C.

Problems 11

1 The circuit shown in Figure 11.26 is used to measure the resistance of components connected between terminals A and B. Determine the value of R_X when the ammeter registers:
(a) 5 A; (b) 50 mA; (c) 25 μA.

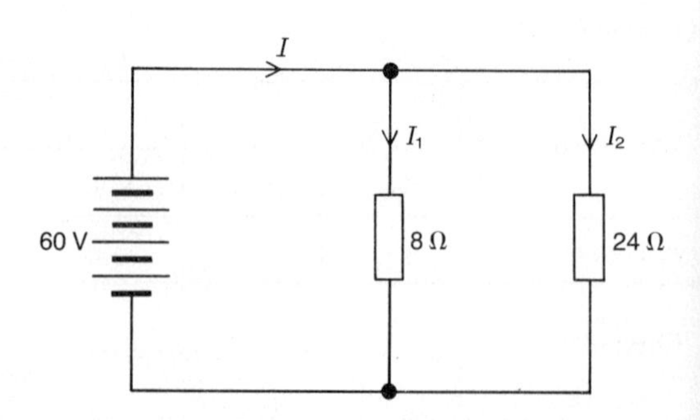

Figure 11.24 For Qu 4

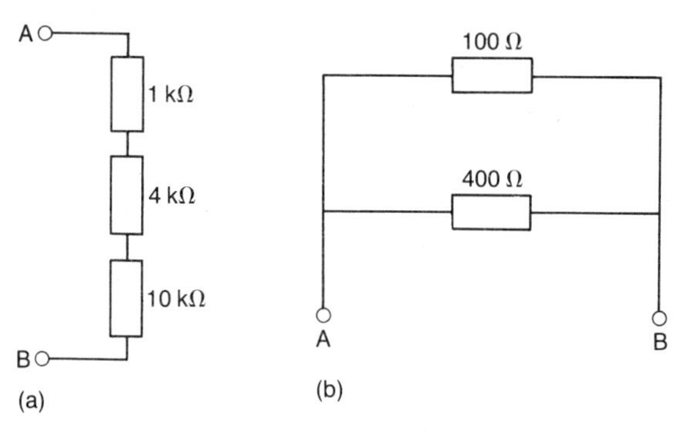

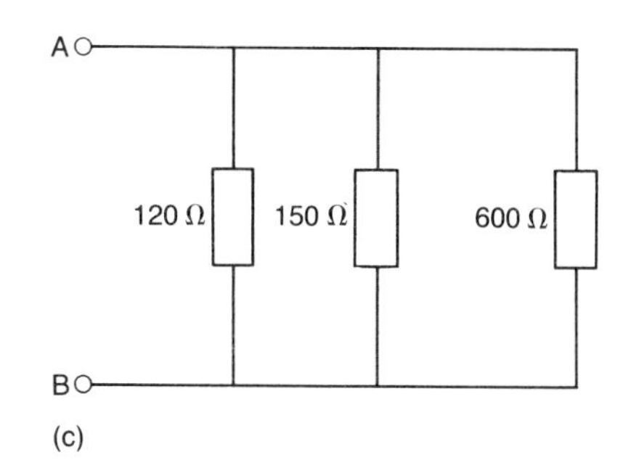

Figure 11.25 For Qu 5

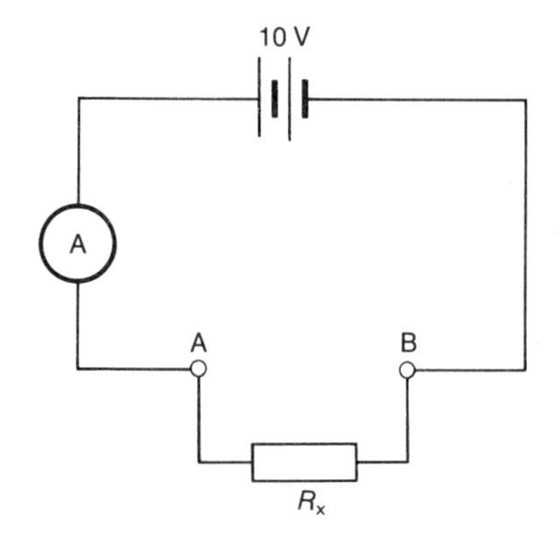

Figure 11.26 For problem 1

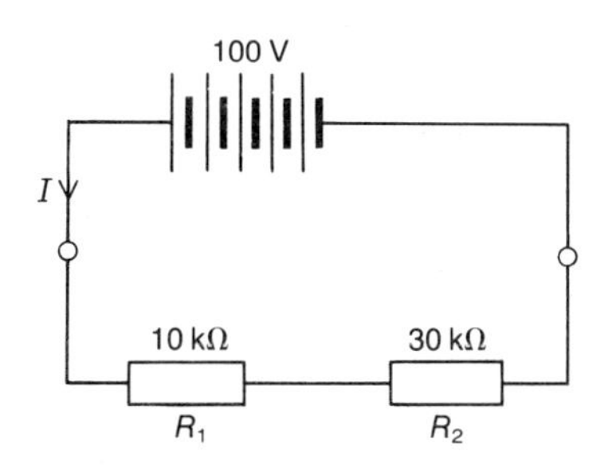

Figure 11.27 For problem 4

2 (a) State Ohm's law. Give two examples of components which conduct electricity and obey Ohm's law and two examples of components which do not satisfy Ohm's law.

(b) Describe an experiment to check the voltage–current relationship for a conductor.

3 Two resistors, $R_1 = 400\,\Omega$ and $R_2 = 600\,\Omega$, are connected in series with a switch, an ammeter and a source of e.m.f. of 20 V. A voltmeter is connected across the 600 Ω resistor.
Draw the circuit diagram and determine what values of current and voltage you would expect to be registered by the meters when the switch is closed.

4 Calculate for the circuit shown in Figure 11.27:
(a) the current I;
(b) the voltage across resistor R_1 and also across R_2.

5 Calculate the equivalent resistance between terminals A–B for the circuit shown in Figure 11.28 and hence calculate the circuit current I.

6 Three resistors 100 Ω, 200 Ω and 300 Ω are connected in parallel and then joined in series with a 45 Ω resistor and a 10 V e.m.f. source. Determine the total resistance of the circuit and the current.

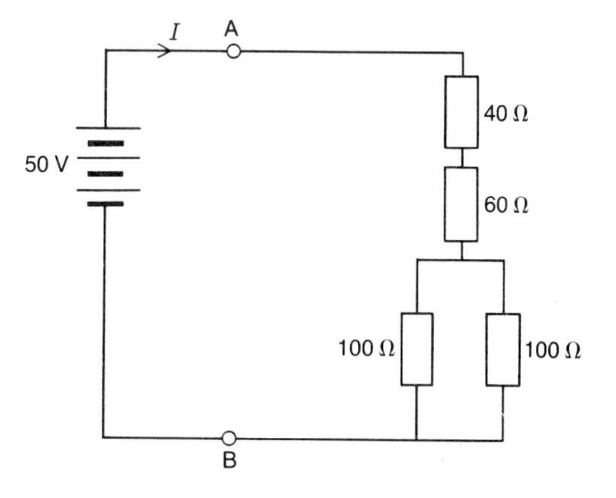

Figure 11.28 For problem 5

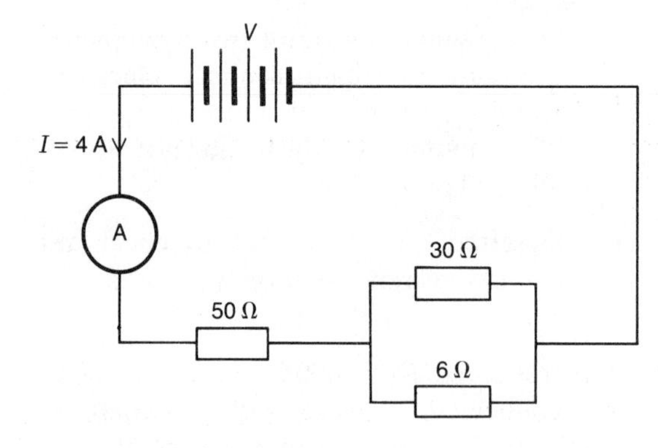

Figure 11.29 For problem 7

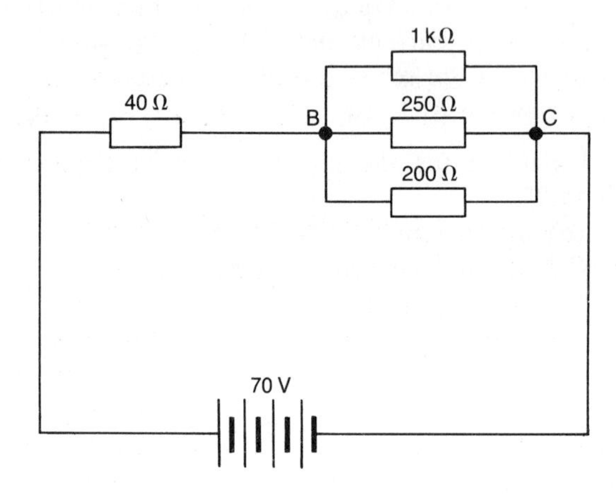

Figure 11.30 For problem 8

7 The current measured by the ammeter in the circuit of Figure 11.29 is $I = 4$ A. Determine:
   (a) the voltage across the 50 Ω resistor;
   (b) the voltage across the 6 Ω and 30 Ω resistors;
   (c) the currents flowing in the 6 Ω and 30 Ω resistors;
   (d) the e.m.f. V of the source.

8 For the circuit shown in Figure 11.30, determine:
   (a) the total equivalent resistance;
   (b) the current supplied by the generator;
   (c) the voltage drop across BC;
   (d) the current in each of the three parallel resistors.

9 Figure 11.31(a) shows a high-resistance circuit. Determine the voltage across the terminals A–B when

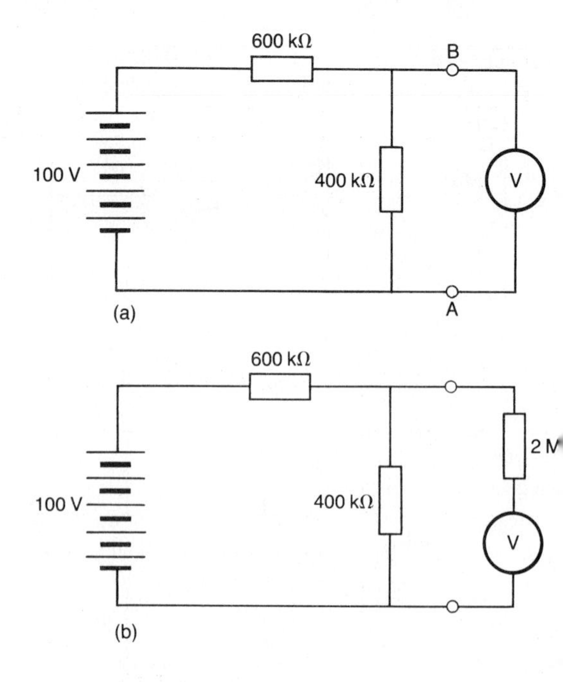

Figure 11.31 For problem 9

   (a) an ideal voltmeter (of infinite resistanc
   is connected across A–B;
   (b) a voltmeter of resistance 2 MΩ is co
   nected across A–B (hint: the effecti
   resistance across A–B is now 400 kΩ i
   parallel with 2 MΩ, see Figure 11.31(b)

10 Define the temperature coefficient of resis
ance of a material and describe an experime
showing how the coefficient could be me
sured over the range 0°C to 100°C.

11 The current measured in a circuit consisting
a 12 V source in series with a resistor and
milliammeter of negligible resistance is 20
mA at 0°C and 148 mA at 50°C. Calculate:
   (a) the resistance of the resistor at 0°C an
   50°C;
   (b) the temperature coefficient of resistanc
   of the material making up the resisto
   over the range 0°C to 50°C.

12 The current in a circuit consisting of a 10
source in series with a resistor is 200 m
when measured at 20°C.

Calculate the current in the circuit whe
the temperature is increased to 32°C if th
temperature coefficient of the resistor ma
erial is (a) 0.004 °C^{-1}; (b) 0.0001 °C^{-1}.

12 Power in electrical circuits

General learning objectives: to calculate power in electrical circuits.

12.1 Power in a resistive circuit: $P = IV = I^2R = V^2/R$

In Chapter 10 we defined the e.m.f. of a source and voltages (potential differences) developed in a circuit in terms of energy and work done:

e.m.f. or source voltage = energy supplied by source in transporting 1 coulomb of charge completely around a circuit.

Voltage between two points in a circuit = work done in transporting 1 coulomb between the two points.

Let us now apply these definitions to find the energy and power supplied by a source to a resistive circuit.

Consider the simple circuit of Figure 12.1 where a source of e.m.f. V is applied across a resistor R. The source e.m.f. will cause a steady flow of charge to be produced in the circuit, that is, it will sustain a dc current. Suppose the value of this is I amperes (coulombs per second).

Then, as

charge = current × time

the charge Q transported around the circuit in a time of t seconds is given by

$Q = It$ coulombs

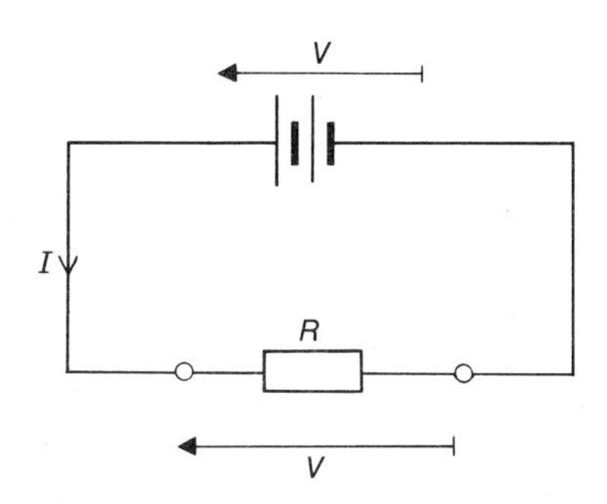

Figure 12.1 Power supplied and dissipated in resistive circuit is, $P = IV = I^2R = V^2/R$ *watts*

and as each coulomb of charge requires the source to supply an energy of V joules (V = energy supplied per coulomb), the total energy supplied by the source in t seconds is given by

$$W = Q \times V = (It) \times V = IVt \text{ joules}$$

Further, since power is the rate of transfer or supply of energy, the power P supplied by the source is

$$P = \frac{W}{t} = IV \text{ joules per second, or watts}$$

The energy supplied by the source to sustain the current is dissipated in the resistor as heat. The energy is used up in doing work to overcome the resistance presented by the conductor(s) of the circuit. Electrical energy supplied by the source of e.m.f. is converted to heat in the resistive elements of a circuit. If the total resistance of a circuit is R ohms, the total current supplied by the source is given by Ohm's law as

$$I = \frac{V}{R}$$

where V = applied e.m.f. = voltage developed across the resistors so we can express the power P supplied to and dissipated in a resistive circuit as

$$P = IV \text{ watts}$$

or $P = I \times (RI) = I^2R$ watts, as $V = RI$

or $P = (V/R) \times V = V^2/R$ watts, as $I = V/R$

12.2 Calculations of energy and power dissipated in resistive circuits

Examples

1 Calculate the power supplied to the circuit of Figure 12.2 and the power dissipated in the 20 Ω and 30 Ω resistors.

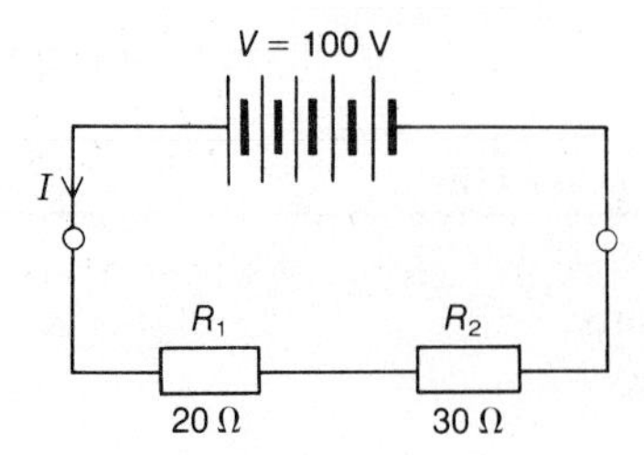

Figure 12.2 Circuit for example 1

Solution

The total resistance of the circuit,

$$R = R_1 + R_2 = 20 + 30 = 50\,\Omega$$

so the circuit current,

$$I = \frac{V}{R} = \frac{100}{50} = 2\,\text{A}$$

Thus the power supplied by the 100 V source to the circuit,

$$P = IV = 2 \times 100 = 200\,\text{W} \quad \textit{Ans}$$

and the power dissipated in the $R_1 = 20\,\Omega$ resistor,

$$P_1 = I^2R_1 = 2^2 \times 20 = 4 \times 20 = 80\,\text{W} \quad \textit{Ans}$$

and the power dissipated in the $R_2 = 30\,\Omega$ resistor,

$$P_2 = I^2R_2 = 2^2 \times 30 = 4 \times 30 = 120\,\text{W} \quad \textit{Ans}$$

Note: $P_1 + P_2 = 80 + 120 = 200\,\text{W} = P$

Thus total power dissipated = power supplied by source, exactly as we should expect.

2 Calculate the power dissipated in resistors $R_1 = 10\,\Omega$ and $R_2 = 5\,\Omega$ in the parallel circuit of Figure 12.3.

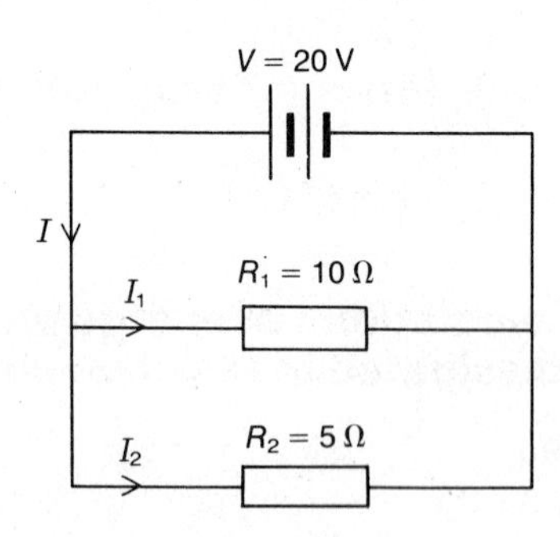

Figure 12.3 Circuit for example 2

Solution

The voltage across both resistors is equal to the source e.m.f. $V = 20\,\text{V}$, so the powers dissipated can be found applying the formula, $P = V^2/R$ directly. Power dissipated in $R_1 = 10\,\Omega$,

$$P_1 = V^2/R_1 = 20^2/10 = 400/10 = 40\,\text{W} \quad \textit{Ans}$$

Power dissipated in $R_2 = 5\,\Omega$,

$$P_2 = V^2/R_2 = 20^2/5 = 400/5 = 80\,\text{W} \quad \textit{Ans}$$

Note: we could, of course, use

$P_1 = I_1^2R_1$, where $I_1 = V/R_1 = 20/10 = 2\,\text{A}$

and $P_2 = I_2^2R_2$, where $I_2 = 20/5 = 4\,\text{A}$

so $P_1 = 2^2 \times 10 = 40\,\text{W}$ and $P_2 = 4^2 \times 5 = 80\,\text{W}$

3 Calculate the current flowing in and the resistance of the heater filament for a car rear-window demister which is rated at 6 W when working off a 12 V battery.

Solution

Let I = current in heater filament when connected to the 12 V battery

and R = resistance of filament,

then as the power dissipated in the filament is 6 W,

we have, $P = IV$

$6 = I \times 12$

so $$I = \frac{6}{12} = 0.5\,\text{A} \quad \textit{Ans}$$

and $$R = \frac{V}{I} = \frac{12}{0.5} = 24\,\Omega \quad \textit{Ans}$$

4 An electric fire takes a current of 12.5 A when connected to a 240 V supply. Calculate:

(a) the power dissipated by the fire, i.e. the heat generated;

(b) the total energy supplied by the source if the fire is switched on for four hours;

(c) the cost of running the fire if the electricity tariff is charged at 7.5p per kilowatt hour.

Note: Although the SI unit of energy is the joule, electrical energy is often quoted in terms of kilowatt hours (kWh),

$$1\,\text{kWh} = 1\,\text{kW supplied for 1 hour}$$
$$= 1000\,\text{W} \times \text{number of seconds in 1 hour}$$
$$= 1000 \times 60 \times 60 = 3.6 \times 10^6\,\text{J}$$

i.e. 1 kWh = 3.6×10^6 joules or 3.6 megajoules (MJ)

Solution

(a) the power dissipated by the fire,

$P = IV$ watts
where $I = 12.5\,\text{A}$, $V = 240\,\text{V}$
so $P = 12.5 \times 240 = 3000\,\text{W}$ or 3 kW *Ans*

(b) The total energy supplied in four hours

$W = P \times t$
where $P = 3\,\text{kW}$ and $t = 4\,\text{h} = 4 \times 60 \times 60 = 14\,400\,\text{s}$
so $W = (3\,\text{kW}) \times (4\,\text{h}) = 12\,\text{kWh}$
or $W = (3000\,\text{W}) \times (14\,400\,\text{s}) = 43.2\,\text{MJ}$ *Ans*

(c) The cost of running the fire,

cost = (number of kilowatt hours) × (cost per killowatt hour)
= 12 × 7.5 = 90p or £0.90 *Ans*

5 Calculate for the circuit of Figure 12.4,
(a) the total circuit resistance R and circuit current I;
(b) the currents I_2 and I_3, flowing in resistors $R_2 = 20\,\Omega$ and $R_3 = 30\,\Omega$;
(c) the powers dissipated in resistors $R_1 = 8\,\Omega$ and R_2 and R_3;
(d) the total power supplied to the circuit.

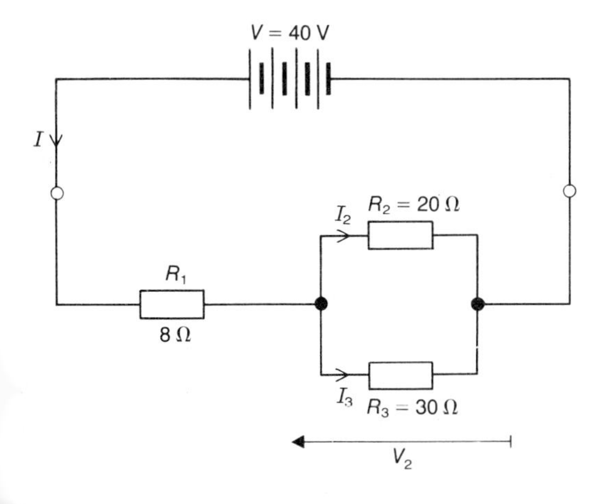

Figure 12.4 Circuit for example 5

Solution

(a) The total resistance of the circuit, $R = R_1$ + parallel combination of R_2 and R_3

Now, $$\frac{1}{R_2} + \frac{1}{R_3} = \frac{1}{20} + \frac{1}{30} = \frac{3+2}{60} = \frac{5}{60}$$

so the resistance of R_2 in parallel with R_3

$$= \frac{60}{5} = 12\,\Omega$$

and hence
$R = R_1 + 12 = 8 + 12 = 20\,\Omega$ *Ans*

and the circuit current,

$$I = \frac{V}{R} = \frac{40}{20} = 2\,A$$ *Ans*

(b) The voltage V_2 across the paralleled resistors,

$$V_2 = I \times (R_2 \text{ in parallel with } R_3) = 2 \times 12 = 24\,\text{V}$$

so current in R_2, $I_2 = \frac{24}{R_2} = \frac{24}{20} = 1.2\,\text{A}$ *Ans*

and current in R_3, $I_3 = \frac{24}{R_3} = \frac{24}{30} = 0.8\,\text{A}$ *Ans*

(c) The powers dissipated:

in R_1: $P_1 = I_1{}^2R_1 = 2^2 \times 8 = 32\,\text{W}$ *Ans*
in R_2: $P_2 = I_2{}^2R_2 = 1.2^2 \times 20 = 28.8\,\text{W}$ *Ans*
in R_3: $P_3 = I_3{}^2R_3 = 0.8^2 \times 30 = 19.2\,\text{W}$ *Ans*

(d) The total power supplied to the circuit,

$P = IV = 2 \times 40 = 80\,\text{W}$ *Ans*

with checks with

$$P = P_1 + P_2 + P_3 = 32 + 28.8 + 19.2 = 80\,\text{W}$$

12.3 Calculation of fuse values given power rating and voltage of an appliance

When current flows in a conductor or resistive element, electrical power is dissipated and the temperature of the element rises. Familiar examples in which the heating effect is put to use are electric fires, irons, kettles, soldering irons and electric water (immersion) heaters where in each case the heater element is designed to produce a given amount of heat for a given supply voltage. The element must, of course, be able to withstand the temperature to which it may be raised and the stresses and strains caused by heating and subsequent cooling.

It is also of utmost importance to protect electrical appliances against taking excessive currents, which could cause permanent damage and even produce fire. Even more important it is essential to protect the user against electric shock if a fault were to occur to make the appliance 'live'.

For these reasons power supplies and individual appliances should always contain fuses. A fuse is a device for protecting electrical circuits by immediately disconnecting the electrical power supply to the circuit if excessive current flows. One of the most commonly employed fuses consists essentially of relatively fine wire or conducting strip made of a fusible material, usually a metal alloy, which melts when the current exceeds a certain value – the fuse rating current. For example, common household fuses are rated at 3 A, 5 A, 13 A and 15 A, while car fuses may be as high as 35 A and fuses in certain electronic equipment can be as low as 50 mA. Examples of fusible wire type fuses, known as cartridge fuses are shown in Figure 12.5.

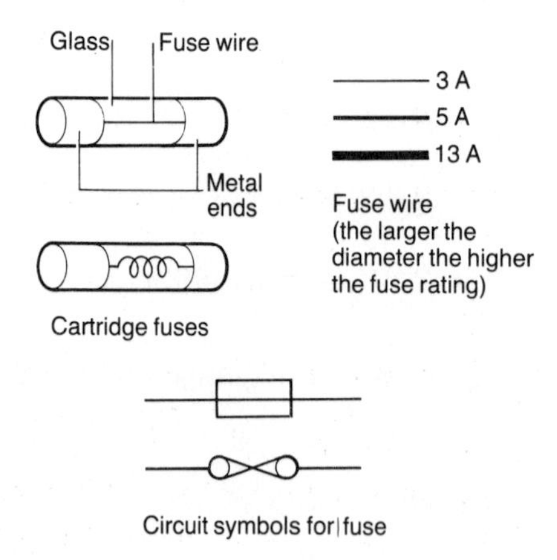

Figure 12.5

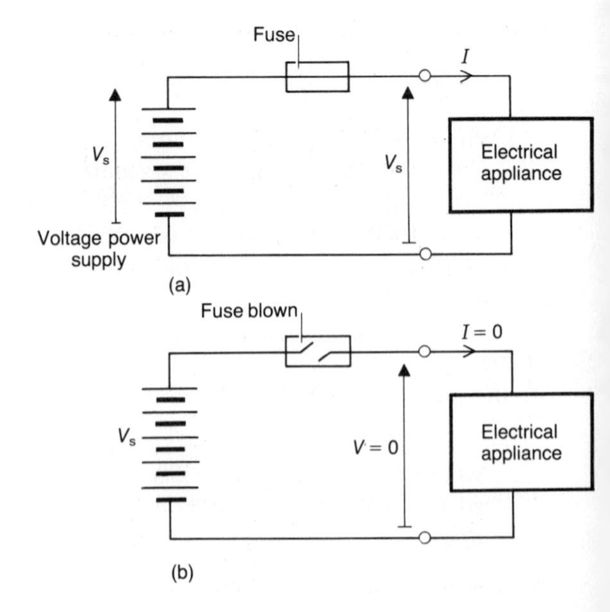

Figure 12.6 Action of a fuse
(a) Circuit operating normally with supply connected to appliance
(b) Fault causes fuse to blow disconnecting appliance from power supply

To protect an electrical appliance the fuse is fitted in series with the electrical power supply as shown in Figure 12.6. Thus if a fault occurs which causes a current to flow in excess of the rated value of the fuse, the fuse will melt and produce a permanent break in the circuit. This break will disconnect the supply from the appliance.

The fuse current rating is determined from the power relation:

$$P = IV$$

where P = power rating of appliance, watts
V = supply voltage, volts
I = current drawn from supply by appliance

Normally the fuse rating is ten to twenty per cent above I to allow for small variations in supply voltage and changes in the effective resistance of the appliance.

A diagram of a typical three-pin mains socket and plug is shown in Figure 12.7. The ac mains

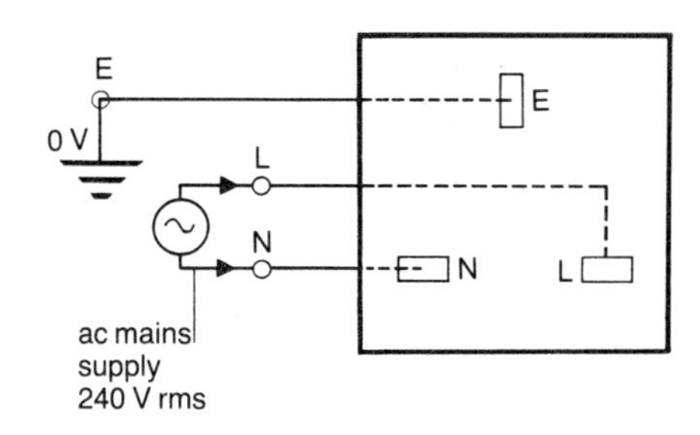

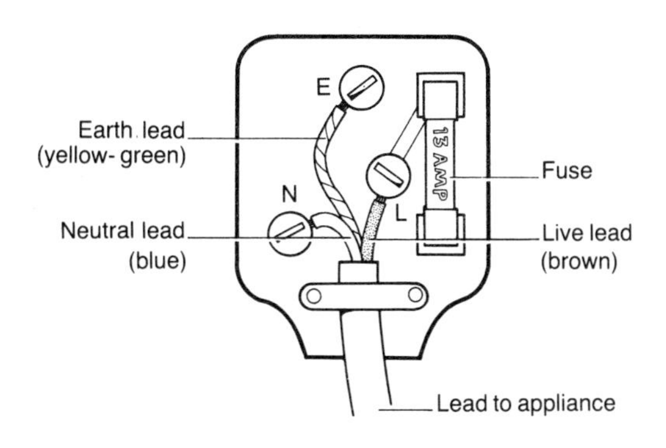

Figure 12.7 3 pin mains socket and plug

voltage is developed across the *'live'* (L) and *'neutral'* (N) terminals. The plug is used to connect our appliance leads to the mains supply. In the UK the ac mains voltage is 240 V [actually this is known as the root mean square (rms) value, the peak value is higher at 340 V, but $V =$ 240 V is the effective value as far as the power $P = IV$ relation is concerned].

The neutral wire has a light blue insulator covering, the live wire a brown one, and the earth wire a yellow-green covering. Note, also, the presence of a fuse adjacent to the live lead terminal in the plug. The fuse forms a series conducting link between the L plug terminal and the L pin on the underside of the plug (not shown on the diagram). Should our appliance take greater than its rated current owing to some fault, this fuse should 'blow' and disconnect the ac supply from the appliance.

The earth wire is also included for reasons of safety, and this wire should be connected to the earth terminal of our appliance. Under correct operating conditions, the earth lead plays no role whatsoever either in connecting the ac mains or in conducting current. However, should a fault occur such that our appliance might become 'live', e.g. if the live wire made a conducting contact with the frame of our appliance, the presence of the earth connection would immediately prevent injury to the person in contact with the appliance. The earth wire holds the frame at zero volts. The earth wire provides a path of extremely low resistance connected directly down to earth at zero volts. Thus, although current may flow for a short time from the live wire down to earth, the ac mains will be rapidly disconnected by the fuse blowing in the plug itself, or by a fuse further back in the mains supply circuits (for example, in the consumer supply unit where the mains electricity supply enters the premises). In addition to or as an alternative to wire fuses most ac electrical supply systems employ safety devices of the earth leakage contact breaker type (ELCB). The ELCB operates by balancing the current in the 'live' line and the return current in the 'neutral' line. If an imbalance occurs this must be due to current in the earth line and such an imbalance causes the breaker to trip, opening the circuit and disconnecting the supply. The ELCB can be reset manually after the fault is cleared.

Examples

1 Determine the fuse current rating which should be used to protect an electrical appliance rated at 3 kW at a supply voltage of 240 V.

Solution

The power drawn from the supply is $P = 3$ kW $= 3000$ W and the supply voltage $V = 240$ V, so using

$$P = IV$$

the current flowing in the appliance is

$$I = \frac{P}{V} = \frac{3000}{240} = 12.5 \text{ A}$$

Thus the fuse current rating must be at least 12.5 A. To allow for minor variations in supply but to protect against fault conditions a sensible rating would be 15 A. It is important to stress that fuses are normally only available in standard current ratings, e.g. for domestic circuits 3 A, 5 A, 13 A, 15 A. It is good practice when calculating a value to round up to the next higher value or to a value, to allow for minor fluctuations in supply or appliance rating, the order of 10 to 20% higher. Thus in this example we could use marginally a 13 A fuse, but to avoid the fuse blowing inadvertantly 15 A.

2 A lighting circuit is protected by a 5 A fuse and is fed by a mains supply voltage of 240 V. The circuit currently supplies eight 60 W and five 100 W light bulbs. Is it advisable to add to this circuit a further five 100 W bulbs?

Solution

The maximum power that the supply can transfer to the lighting circuit without blowing the fuse is,

$$P = IV$$

where $I = 5$ A the fuse rating and $V = 240$ the supply voltage

so $P = 5 \times 240 = 1200$ W

The power consumed by

eight 60 W bulbs $= 8 \times 60 = 480$ W
five 100 W bulbs $= 5 \times 100 = 500$ W

so a total power of $480 + 500 = 980$ W is required. A further five 100 W bulbs would demand 500 W and so the total power drawn from the supply would be $980 + 500 = 1480$ W clearly exceeding $P = 1200$ W. Thus it is not advisable to add the further five bulbs. A new separately fused circuit should be installed.

Test 12

This test may be used as a basic self-assessment test to check your understanding of Chapter 12 on **Power in electrical circuits**, and its learning objectives. Enter all answers in the answer block.

Answer block:

Question no.	1			2			3		
	(a)	(b)	(c)	(a)	(b)	(c)	(a)	(b)	(c)
Answer									

Qu. 1 Enter a tick (√) in the answer block if you consider the statement is correct; enter a cross (×) if you consider the statement is in any way incorrect.
(a) The power supplied to a circuit is given by IV watts where V = supply voltage and I = circuit current.
(b) The power dissipated in a resistor is given by I^2R where R = resistance in ohms and I = current in resistor in amperes.
(c) A fuse is a device for protecting an electrical appliance or circuit. It disconnects the supply if excessive conditions such as high current occur.

Qu. 2 Determine the current that flows in the following devices:
(a) a 60 W light bulb from a supply of 240 V;
(b) a 6 W soldering iron from a 12 V supply;
(c) a 2 kW electric kettle from a 240 V supply.

Qu. 3 A one-bar electric fire is rated at 1 kW when connected to a 250 V supply. Calculate:
(a) the effective resistance of its heating element;
(b) the total energy in joules consumed by the fire in five hours;
(c) the cost of this energy if the tariff is 10p per kilowatt-hour.

Problems 12

1 Show that the power dissipated in a resistor of resistance R is given by $P = I^2R$ where I = current flowing in the resistor.
For the circuit of Figure 12.8 determine

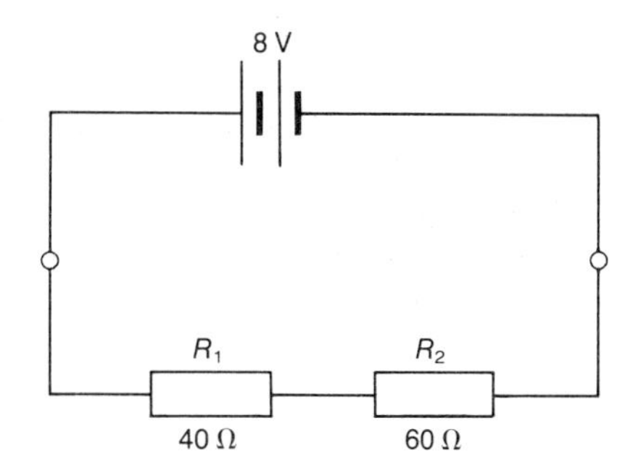

Figure 12.8 For problem 1

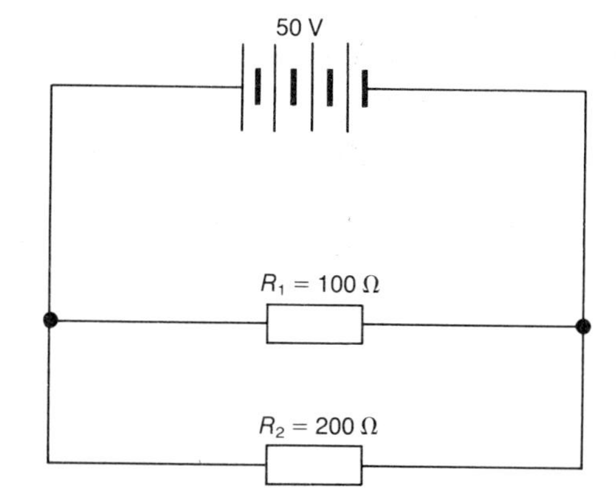

igure 12.9 For problem 2

(a) the circuit current
(b) the power dissipated in resistor $R_2 = 60\ \Omega$
(c) the power supplied to the circuit.

For the circuit of Figure 12.9 determine
(a) the power dissipated in both R_1 and R_2
(b) the total power supplied by the source.

For the circuit of Figure 12.10 determine

(a) the circuit resistance and supply current;
(b) the power dissipated in the 30 Ω resistor.

Determine which of the available fuses you would select to protect the following appliances:
(a) a 10 W, 30 V soldering iron;
(b) a lighting circuit supplying up to 500 W from a supply of 240 V;
(c) a 3 kW immersion heater fed by a 240 V supply.

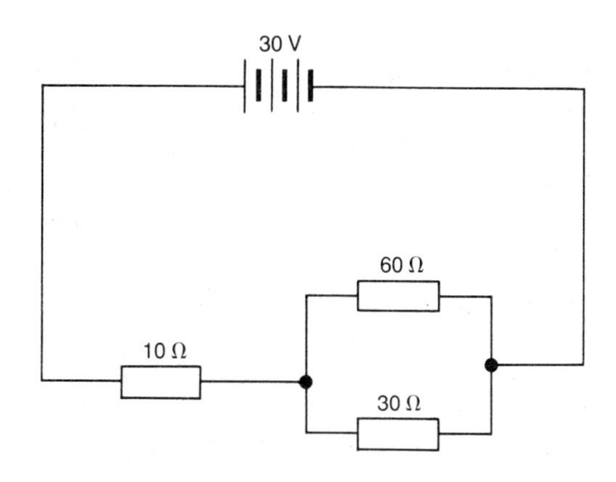

Figure 12.10 For problem 3

Fuses available: 100 mA, 1 A, 3 A, 5 A, 13 A, 15 A, 30 A.

5 Describe an experiment you would perform to test the fuse rating currents for a number of different fuses.

13 Electrochemical effects and applications

General learning objectives: first to consider the conduction process in solids and liquids and then to explain the chemical effects of electricity and its applications.

13.1 Good and bad conductors of electricity: conductors, semiconductors and insulators

In the solid state the atoms making up a substance are closely packed together and in the case of many materials the atoms are structured in regular crystalline patterns. When atoms form a regular crystalline structure, as in metals and graphite, the outermost electrons of the atoms are weakly bound and may easily break away from the atoms. They become 'free' electrons and these negatively charged elements are the charge carriers for electrical current in solids. When acted upon by an electrical force – an applied e.m.f. – a uniform drift is superimposed on their previous random motion. This drift of negatively charged electrons constitutes the electrical current. Metals and carbon graphite are good conductors of electricity since free electrons are readily available.

In an insulating material virtually all electrons are tightly bound to their atoms. Under 'normal' values of voltage no 'free' electrons are produced and hence there is no conduction of current. It requires immense force to free electrons. Several hundred kilovolts applied to an insulator may cause breakdown where electrons are 'torn' from their bound states to conduct electricity. However, under normal circumstances insulators such as ceramics and plastics contain no free electrons and hence do not conduct current.

Semiconductor materials, such as silicon, germanium and gallium arsenide and which are used to fabricate transistors and integrated circuits, have rather special properties. In their pure state they are poor conductors of electricity but have the interesting property that their resistance normally decreases with temperature rise in contrast to metals where the resistance rises with temperature, as we have already seen.

The measure of the ability of a material to conduct electricity is quantified by its **resistivity**. Resistivity, usually denoted by the Greek letter ρ (rho), is the resistance between the opposite faces of a one metre cube of material, see Figure 13.1. Values of resistivity for good conductors, semiconductors and insulators are given in Table 13.1. You can see the immense range of values.

Copper, followed by aluminium, is the most widely used metal for the conduction of electricity for both industrial and domestic applications. Copper has a lower resistance for a given cross-section while aluminium has a lower resistance for a given weight. Increased mechanical strength may be obtained by using composite conducting materials. For example, steel-cored aluminium and copper-clad steel are used, the steel supplying the strength and the aluminium and copper providing the good electrical conducting properties. Such properties are required by overhead lines used to distribute electricity. These conducting lines must have sufficient mechanical strength to hang freely in a fairly long span and save the number of pylons required.

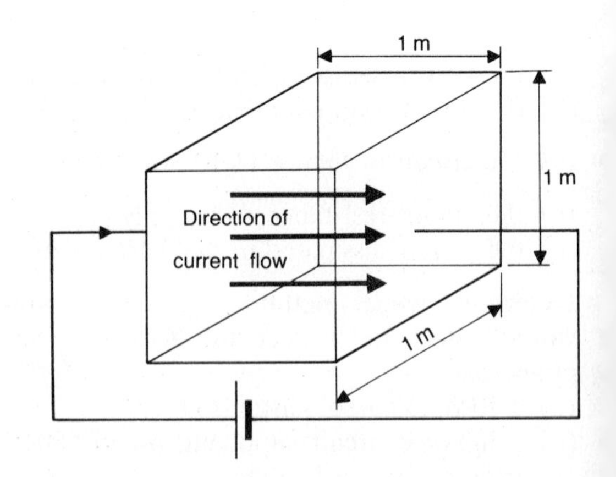

Figure 13.1 Resistivity = resistance of metre cube material

Table 13.1 Resistivity for conductors, semiconductors and insulators

Material	*Resistivity* ρ (ohm-metres at 20°C)	*Relative resistance* = ρ *material* / ρ *copper*
copper	1.72×10^{-8}	1
aluminium	2.8×10^{-8}	1.64
iron	9.6×10^{-8}	5.6
platinum	10.6×10^{-8}	6.16
silver	1.6×10^{-8}	0.95
gold	2.4×10^{-8}	1.42
steel	12 to 90×10^{-8}	7 to 53
carbon	$\sim 7000 \times 10^{-8}$	~4000
manganin	45×10^{-8}	26
constantan	49×10^{-8}	28.5
pure silicon	2.3×10^{3} (at 27°C)	
pure germanium	0.5 (at 27°C)	
ceramics	$>10^{12}$	
porcelain	$>10^{11}$	
glass	$>10^{7}$	
polythene	$>10^{15}$	
paper	$>10^{9}$	
ubbers	$>10^{11}$	

They must also possess low resistance to carrying full current without undue heat loss or tempera-ure rise. Forms of carbon are also used to conduct electricity, for example, carbon brushes n an electrical motor and carbon resistors. Precious materials, such as platinum and gold, are ometimes used for specialist applications where a metal such as copper may deteriorate through oxidation.

Insulating materials do not conduct electricity nd are therefore used for covering or separating onductors carrying current, or to separate onductors at different voltages to prevent urrent flow between the conductors. For exam-le, ceramic materials are used for constructing sulators to carry conductors for electrical supply nd lines for telecommunication applications. lastics are used widely as insulators in electrical ppliances and electronic equipment. Plastic aterial, such as polythene, is also used to eparate conductors in wiring cable (for example, two- and three-wire lighting and power cable) nd coaxial cables for telecommunications pur-oses. An example of the latter is the coaxial able which joins an aerial to a television set. The inner central wire is separated from the outer conducting sheath by polythene.

13.2 Conduction of electricity in liquids and its chemical decomposition effects: electrolytes and electrolysis

Many crystalline substances do not conduct electricity in the solid state but readily conduct in the molten (liquid) state and when they are dissolved in a solvent such as water. The conduction of electricity in the molten state or when in solution is due to the formation of **ions**. Ions are charged atoms (atoms which have gained or lost one or more electrons) and also groups of charged atoms or charged parts of a molecule. For example, common salt (sodium chloride, NaCl) does not conduct in the solid state. When it is dissolved in water, sodium ions Na^+ and chlorine ions Cl^- are formed; the (+) attachment denotes that the sodium atom Na^+ has lost one electron and therefore is positively charged, the (−) attachment indicates that the chlorine atom has gained one electron and is therefore negatively charged. In the solid state the ions are held firmly in place in a crystalline array by attraction between the ionic charges. When the salt is dissolved in water the attraction forces are considerably weakened and many ions dissociate and take up a free existence and are available to act as charge current for the conduction of current.

Chemical compounds, which in their molten state or when dissolved in a solvent form 'free' ions and therefore can conduct electricity, are known as **electrolytes**. Solutions of most acids (e.g. sulphuric acid H_2SO_4, hydrochloric acid HCl), salts (e.g. zinc sulphate $ZnSO_4$, silver nitrate $AgNO_3$) and bases (e.g. sodium hydroxide NaOH) are examples of electrolytes: they form free ions and are conductors of electricity.

If no free ions are readily formed the solution cannot conduct electricity. For example, pure distilled water contains few ions and is a very poor conductor – its resistivity is between 10^2 and 10^6 Ωm. Oil contains even fewer ions and is often used as an insulator, e.g. silicone oil has a resistivity of 10^{12} Ωm, comparable with good solid insulator values.

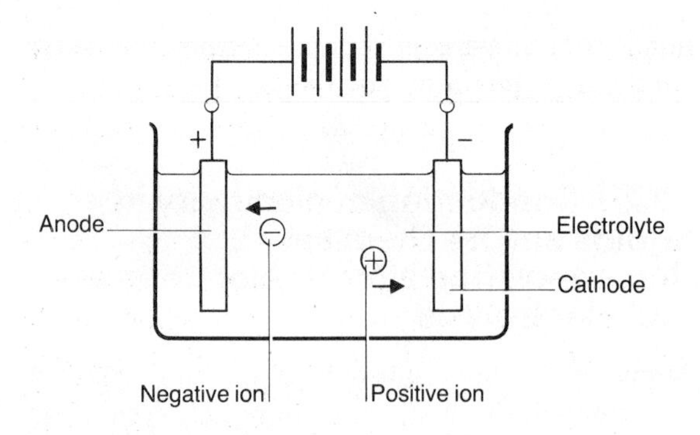

Figure 13.2

When electric current is caused to flow in an electrolyte it produces a chemical effect and this effect is known as **electrolysis**. The main chemical action is observed normally at and close to the terminal conductors at which current enters and leaves the electrolyte. The electrode connected to the positive side of the voltage source and at which current enters the electrolyte is known as the **anode** and the electrode connected to the negative polarity side of the source and at which current leaves the electrolyte is known as the **cathode**, see Figure 13.2.

When the voltage source is applied across the electrodes the positive ⊕ ions in the electrolyte are attracted to the cathode and the ⊖ negative ions to the anode. As they reach the electrodes they give up their charge and become uncharged atoms or groups and are either liberated or deposited at the electrode or react chemically with the electrode or solvent.

Let us take a specific example of electrolysis and also consider the mechanism of current flow between two platinum electrodes in a solution of zinc chloride ($ZnCl_2$) electrolyte (see Figure 13.3).

1 Present in the solution are Zn^{++} ions (zinc atoms having lost two electrons) and Cl^- ions (chlorine atoms having gained one electron).
2 When switch S is closed current flows in the circuit. In the solution Zn^{++} ions are attracted to the cathode (the platinum plate connected to the negative terminal of the battery source) and Cl^- ions are attracted to the anode (platinum plate connected to the positive terminal of the battery).
3 When a Zn^{++} ion reaches the cathode it receives two electrons drawn from the voltage supply and is converted into a zinc atom. Thus zinc is deposited on the cathode.
4 When a Cl^- ion reaches the anode it gives up its electron and combines with another chlorine atom (simultaneously or previously neutralized at the anode) to form a molecule of chlorine. Chlorine gas is liberated at the anode and the electronic charges given up are returned to the voltage source.
5 Thus we can say that the mechanism of current flow in the solution is due to zinc and chlorine ions and that the electrolysis has liberated chlorine at the anode and caused zinc to be deposited at the anode, showing chemical decomposition has indeed taken place.

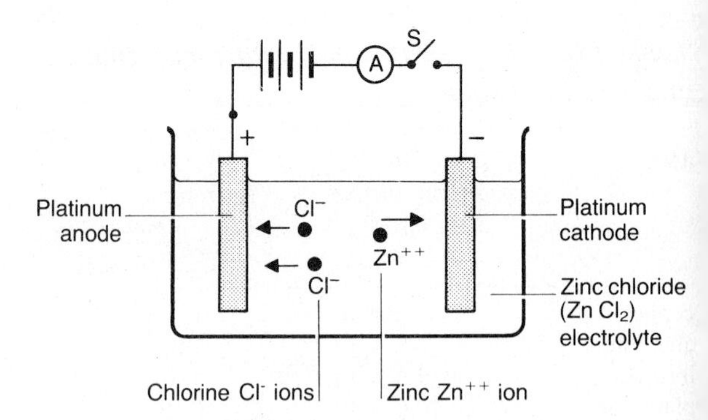

Figure 13.3 Conduction of zinc chloride solution: electrolysis liberates chlorine gas at anode and deposits zinc on cathode

13.3 Electrodeposition: electroplating and electro-refining

Electrolysis – the chemical decomposition o electrolytes by electric current – has man important industrial applications, the two mai ones being electroplating and electro-refining.

Electroplating is the process whereby a layer c metal is deposited on an object by means c electrolysis. The object to be plated is made th cathode, i.e. it is connected to the negativ terminal of a battery or the dc voltage sourc used. The anode, the electrode connected to th positive voltage source terminal, is made of th plating metal. The electrolyte consists of th plating metal, usually a salt, dissolved in suitable solvent such as water.

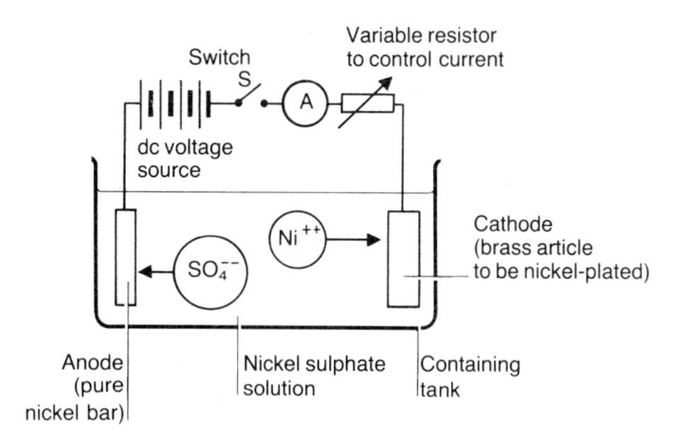

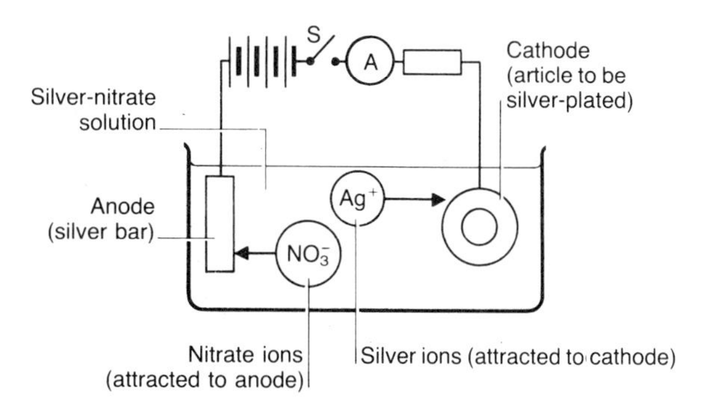

Figure 13.4 Diagrams illustrating two electroplating processes

Two examples of electroplating: nickel and silver plating may be described with reference to Figure 13.4. In (a) the anode consists of nickel (the plating material), the electrolyte is a solution of nickel sulphate (containing nickel Ni^{++} and sulphate SO_4^{--} ions) and the material to be plated (e.g. brass) is made the cathode. When switch S is closed the Ni^{++} ions are attracted to the cathode where their charge is neutralised by each ion gaining two electrons from the supply. The resulting nickel atoms adhere to the cathode and as the current continues to flow layers of nickel build up. At the same time the SO_4^{--} ions are attracted to the anode where they give up their charge to the supply and react with the nickel anode material forming nickel sulphate which returns to the electrolyte solution. Thus nickel is eaten away and puts back an equal amount to that lost from the electrolyte by deposition at the cathode.

Note that the electric current in the electrolyte is carried by the Ni^{++} and SO_4^{--} ions and the current in the supply circuit is electrons. Each time a Ni^{++} ion reaches the cathode it draws two electrons from the supply and each time a SO_4^{--} ion reaches the anode it delivers two electrons back to the supply.

The silver plating process of Figure 13.4(b) is exactly similar. The anode is silver, the electrolyte silver nitrate ($AgNO_3$ containing silver Ag^+ and nitrate NO_3^- ions) and the cathode is the object to be plated. When current flows, Ag^+ ions are attracted to the cathode where they are neutralized and deposited. NO_3^- ions are attracted to the anode where they give up an electron and react with the silver anode to produce silver nitrate, thus replenishing the silver taken out of the electrolyte by deposition at the cathode.

Michael Faraday (1791–1867) investigated electrolysis and found that the mass of substance deposited or liberated at an electrode was directly proportional to the quantity of electricity passed (i.e. the quantity of charge). Thus as,

$$\text{charge} = \text{current} \times \text{time} = It \text{ coulombs}$$

where I = current flowing in electrolyte, A
t = time in seconds

the mass of metal deposited and hence the plating thickness can be controlled by adjusting the current and the time for which it flows. To achieve a good tenacious deposit the current density is normally limited to below 250 A/m^2.

The mass of material deposited or liberated by one coulomb of charge is known as the electrochemical equivalent of the ion. Some values are given in Table 13.2 below. Thus, if we denote the electrochemical equivalent by z, the mass m deposited or liberated is given by

$$m = zIt \text{ kilograms}$$

Table 13.2 Electrochemical equivalents of some metal ions

Ion	z kilograms per coulomb
Aluminium, Al^{+++}	9.326×10^{-8}
Silver, Ag^+	1.118×10^{-6}
Copper, Cu^{++}	3.294×10^{-7}
Nickel, Ni^{++}	3.041×10^{-7}
Gold, Au^+, Au^{+++}	2.04×10^{-6}, 0.68×10^{-6}

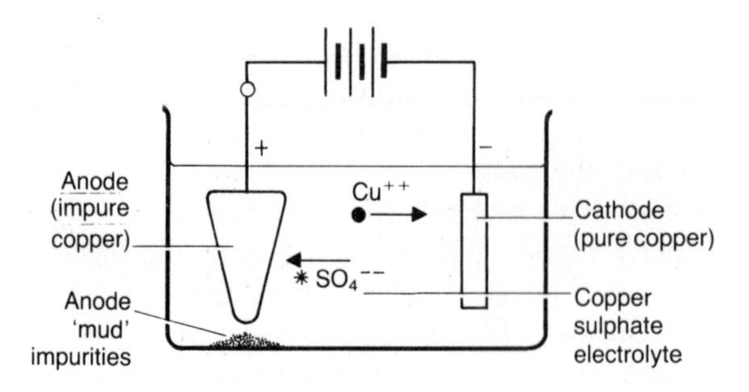

Figure 13.5 Electro-refining of copper

Electrolysis is also used widely in refining certain metals and in the production of metals from their ores. Figure 13.5 shows the principle of electro-refining copper. A large anode of impure copper and a smaller cathode of pure copper are immersed in copper sulphate solution (electrolyte containing copper Cu^{++} ions and sulphate SO_4^{--} ions). The supply voltage is adjusted so that sufficient current flows for SO_4^{--} ions to just react with the copper in the anode and not any more valuable impurities such as silver or gold which fall to the bottom of the vessel and form anode 'mud'. Cu^{++} ions are attracted to the pure copper cathode and in this way 99.96% pure copper may be obtained from the impure anode.

Figure 13.6 shows a diagram illustrating how aluminium may be obtained by electrolysis of its naturally occurring aluminium oxide (bauxite) ore. The aluminium oxide is dissolved in molten cryolite at 1000°C to form an electrolyte containing aluminium Al^{+++} and oxygen O^{--} ions. Carbon electrodes are used, with the Al^{+++} ions being attracted to the cathode and oxygen ions to the anode. The latter react with the carbon to produce oxides of carbon which are liberated and also eat away the anodes. Aluminium in its molten state is collected at the cathode forming the lining of the cell and is drawn off at the bottom and cast into slabs.

Examples

1 Given that the electrochemical equivalent of silver is 1.118×10^{-6} kilograms per coulomb, calculate the mass of silver deposited on the cathode in the electrolysis of silver nitrate (see Figure 13.4(a)) if a current of 5 A is maintained for three hours.

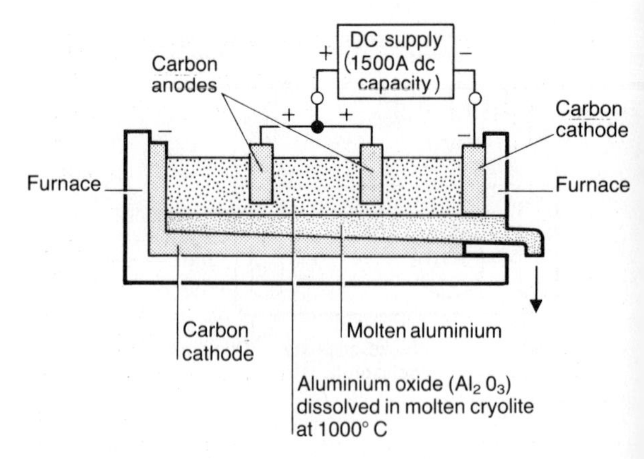

Figure 13.6 The Hall process to produce aluminium from aluminium oxide ore

Solution

The mass of silver deposited

$$m = zIt$$

where $z = 1.118 \times 10^{-6}$ kg/C

$I = 5$ A

and $t = 3\text{ h} = 3 \times 60 \times 60 = 1.08 \times 10^4$ s

hence $m = 1.118 \times 10^{-6} \times 5 \times 1.08 \times 10^4 = 6.04 \times 10^{-2}$ kg or 60.4 g *Ans*

2 Determine the mass of aluminium produced per minute in the Hall process (see Figure 13.6) when a current of 1500 A is passed through the cell. The electrochemical equivalent of aluminium is 9.326×10^{-8} kg/C.

Solution

Mass of aluminium produced,

$$m = zIt$$

where $z = 9.326 \times 10^{-8}$, $I = 1500$ A and $t = 60$ s,

so $m = 9.326 \times 10^{-8} \times 1500 \times 60 = 8.39 \times 10^{-3}$ kg or 8.39 g *Ans*

13.4 Simple cells: principle and construction

A simple cell stores chemical energy which may be converted to electrical energy. The cell acts as a voltage source and may be used to drive electric current around an external circuit.

A cell in its simplest form is made by placing two plates or rods (the electrodes) of different materials in an electrolyte. The electrode materials are normally metals or a metal and carbon. The electrolyte is normally a solution of an acid, base, or salt, in water. Such cells are known as 'wet' cells. In a 'dry' cell the electrolyte is a water-moist paste rather than a liquid solution. When the electrodes are in the electrolyte a chemical reaction occurs between the individual electrodes and electrolyte. The result of these two reactions (one at each electrode) is that each electrode either gains or loses charge and hence a potential difference, known as an electrode potential, is established between each electrode and the electrolyte. The difference between the electrode potentials is equal to the e.m.f. of the cell. The e.m.f. of a cell depends on the nature of the electrodes and electrolyte used, on the concentration of the electrolyte, and on temperature.

Consider, for example, the simple cell shown in Figure 13.7. The cell consists of a copper and a zinc electrode in a solution of dilute sulphuric acid. H^+ ions from the sulphuric acid electrolyte tend to accumulate on the copper electrode making it positively charged. At the zinc electrode Zn^{++} ions tend to enter the electrolyte leaving the zinc electrode negatively charged. Thus we have the state shown in (a) with the copper electrode at a higher voltage with respect to the electrolyte and the zinc electrode at a lower voltage. The difference between the two electrode voltages is equal to the e.m.f. of the cell. For example, if the electrode potential of the copper electrode were +0.5 V and that of the zinc −0.5 V, the cell e.m.f. = 0.5 − (−0.5) = 1 V.

When a resistor is connected across the two electrodes, as shown in (b), the electrons left behind on the zinc plate by the zinc ions entering the solution can now flow through the resistor to the copper plate and neutralize the positively charged hydrogen ions accumulated there. For every two H^+ ions neutralized, one Zn^{++} ion enters into solution from the zinc electrode. Each of these Zn^{++} ions displaces in turn two more H^+ ions from the solution on to the copper plate. Thus an electron current is established between the zinc electrode via the resistor to the copper electrode. Since we regard by convention the

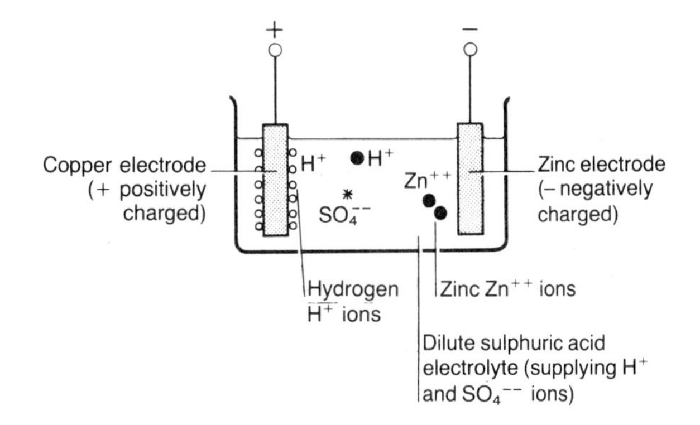

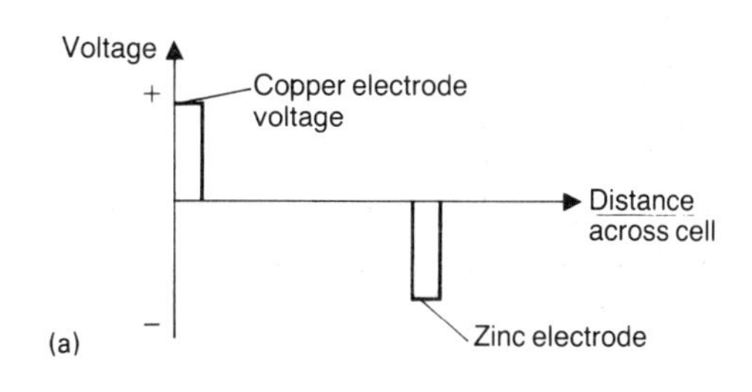

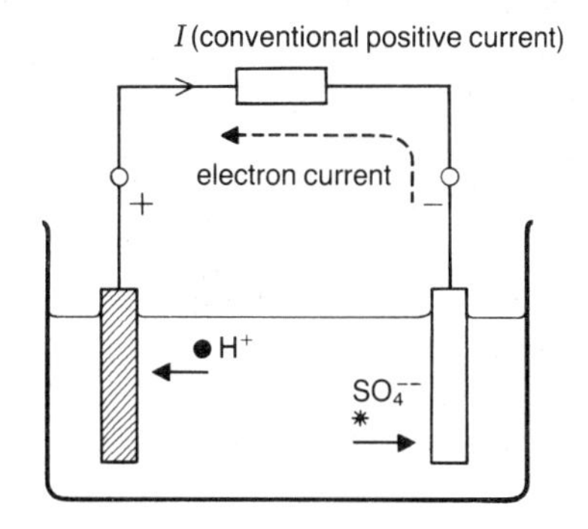

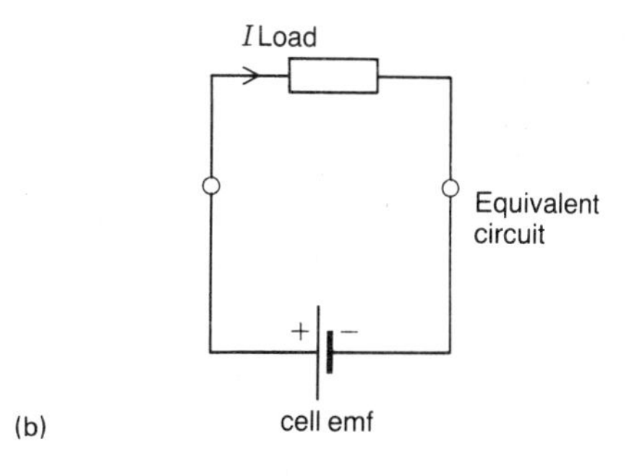

Figure 13.7 Example of a simple cell
(a) Copper-zinc electrodes in sulphuric acid electrolyte forming a simple cell
(b) Simple cell on load, the e.m.f. generated by the cell driving current in an external load

direction of current flow as the direction in which positive charge moves, that is opposite in direction to the flow of electrons, we can say that the electron current is equivalent to a conventional current flow from copper via the resistor to zinc. The copper electrode acts as the positive and the zinc electrode as the negative terminal of the cell when the cell acts as a voltage source.

Two very important points should also be noted. The action at the zinc electrode results in the loss of zinc and hence eventually this electrode will be destroyed. Hydrogen gas is formed at the copper electrode by neutralization of hydrogen ions. This layer of gas impedes current flow and if allowed to build up will greatly increase the internal resistance of the cell and finally render it useless as a source.

The erosion of an electrode in a cell can in certain circumstances be reversed by the process of charging (see next section) but in many practical battery cells the erosion process finally ends the cell's useful life. The presence of gas at the other electrode may be reduced in practice by surrounding the electrode by a chemical 'depolarizing' agent, which oxidizes the gas thereby removing most of the gas formed.

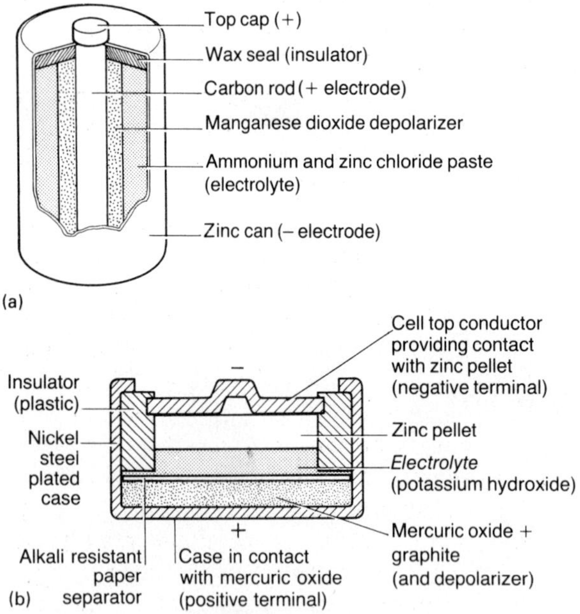

Figure 13.8 Diagrams showing basic construction of two widely used primary cells
(a) Leclanché dry cell: e.m.f. 1.5 V
(b) Mercury (Mallory–Ruben) cell: e.m.f. 1.2 V

The construction of two very widely used cells, the dry Leclanché cell and the mercury cell, is shown in Figure 13.8. The Leclanché cell, named after Georges Leclanché (1839–82), is the mainstay for relatively inexpensive dc batteries. The e.m.f. of a single cell is 1.5 V and the cell is capable of delivering up to several hundred milliamperes for limited periods. The positive electrode is usually in the form of a carbon rod and is surrounded by manganese dioxide to act as a depolarizing agent to remove hydrogen gas formed in use. A paste consisting of ammonium and zinc chloride is used as the electrolyte and a zinc can acts as the negative electrode.

The mercury cell was developed originally by Ruben and Mallory and hence it often bears their names in its description. The cells have a high ratio of energy to weight and volume and can be readily made in miniature form. For example, a common type of mercury cell of button construction similar to that shown in the figure has a diameter of 16 mm and is 16 mm high. It has an e.m.f. of about 1.2 V when supplying 30 mA of load current, and a capacity of one ampere-hour (1 Ah) at this current.

The negative electrode of the cell is formed from an amalgamated zinc powder pressed into pellets. The positive electrode is a mixture of mercuric oxide and graphite, the graphite being added to improve conductivity. The electrolyte is normally potassium hydroxide. The battery casing is in contact with the mercuric oxide but electrically insulated from the other parts of the cell and acts as the positive terminal of the battery. The zinc pellet negative electrode is connected to the top conducting cap which acts as the negative source terminal.

13.5 The difference between primary and secondary cells

The production of an e.m.f. by chemical action by means of inserting two electrodes in an electrolyte leads to two main types of cell: primary cells and secondary cells.

A primary cell is one which uses up its active material during the time that it supplies current to a circuit. Eventually, most of the active material will be used up and then the cell can no longer give a useful supply of voltage. At this stage either the active material must be renewed or the cell must be thrown away. The dry battery used in torches, portable radios, etc. is an example of a primary cell.

A secondary cell is one in which the chemical action is reversible. When a cell supplies current, active material is used up, just as in the case of a primary cell. However, the active material may be subsequently restored by passing a current through the cell in the opposite direction to that of the current normally supplied by the cell. This process, known as charging, reverses the previous chemical process which took place when the cell was supplying current. The most common example of a secondary cell is the car battery. When we start a car, we draw a considerable amount of current from the battery. However, as soon as the engine is running, an electromechanical generator (e.g. a dynamo or alternator) charges the battery, thus reversing the previous chemical action and restoring the active material in the battery.

Examples of practically used primary and secondary battery cells are listed in Table 13.3 below.

13.6 Secondary cells: charging, discharging and the lead–acid accumulator

One of the most widely used secondary cells is the lead–acid cell, which in its most basic form can be constructed by inserting two lead plates into a solution of dilute sulphuric acid as shown in Figure 13.9(a).

Initially no voltage difference is present across these lead electrodes. The cell must first be charged from an external voltage source. Figure 13.9(b) shows a simple charging circuit. The electrode connected to the positive terminal, higher-voltage side of the source, will become the positive terminal of the secondary cell when charged. The electrode connected to the lower-voltage side of the source will become the negative terminal of the cell.

During the charging process, the positive plate (the anode) is coated with a film of active material

Table 13.3 Primary and secondary battery cells

Type	+ *Electrode*	− *Electrode*	*Electrolyte*
Primary			
Leclanché	Carbon	Zinc	Ammonium chloride
Mallory	Mercury	Zinc	
Secondary			
Lead–acid	Lead peroxide	Lead	Sulphuric acid
Nickel–iron	Nickel + hydroxide	Iron + mercury	Potassium hydroxide
Nickel–cadmium	Nickel hydroxide	Cadmium	Potassium hydroxide

and this material is responsible for establishing an e.m.f. in the cell. The charging current in the electrolyte is due to positive hydrogen and negative sulphate ions. The sulphate ions are attracted towards the positive plate, where they give up their charge. Oxygen, which is liberated when the ions give up their charge, then reacts with the lead and forms a brown coating of lead peroxide – the active material. The hydrogen ions are attracted to the negative plate (the cathode), where they give up their charge. Hydrogen gas is thus liberated at the cathode, but no chemical action occurs at this plate.

After this first charging process, the cell has an e.m.f. and could be used as a voltage source in its own right and supply current to an external load, as shown in Figure 13.9(c), although a simple cell of this type, with just a single charge, would be unable to sustain a current for very long. When the cell supplies current, we say that the cell is undergoing discharge. However, if the process of charge followed by discharge is repeated, the capacity of the cell to supply current is steadily increased, since more and more lead at the positive plate is transformed into the active lead peroxide material. In practice, this method of continual charge and discharge is employed only in the production of accumulator plates; an accumulator is the term used to describe a storage battery composed of secondary cells. A practical lead–acid accumulator would already have active material deposited on its anode and would require only a single charge before it could be used.

When the cell discharges, the lead peroxide on the positive plate is gradually changed into a coating of lead sulphate and water. Thus, the sulphuric acid electrolyte becomes diluted and hence its relative density falls. In fact, the state of charge or discharge of a cell is measured by noting the relative density of its sulphuric acid electrolyte. When fully charged, the relative density is about 1.26 and the cell e.m.f. is 2.1 V. Actually, the e.m.f. can rise to about 2.6 V on the final stage of charge, although on discharge the e.m.f. will fall very rapidly to 2.1 V. The e.m.f. of the cell falls on discharge, and in practice should not be allowed to drop below 1.8 V, when the relative density of the electrolyte is approximately 1.17. The discharge curve (terminal voltage versus time) of a lead–acid cell is shown Figure 13.9(d). During discharge, lead sulphate is also formed on the cathode.

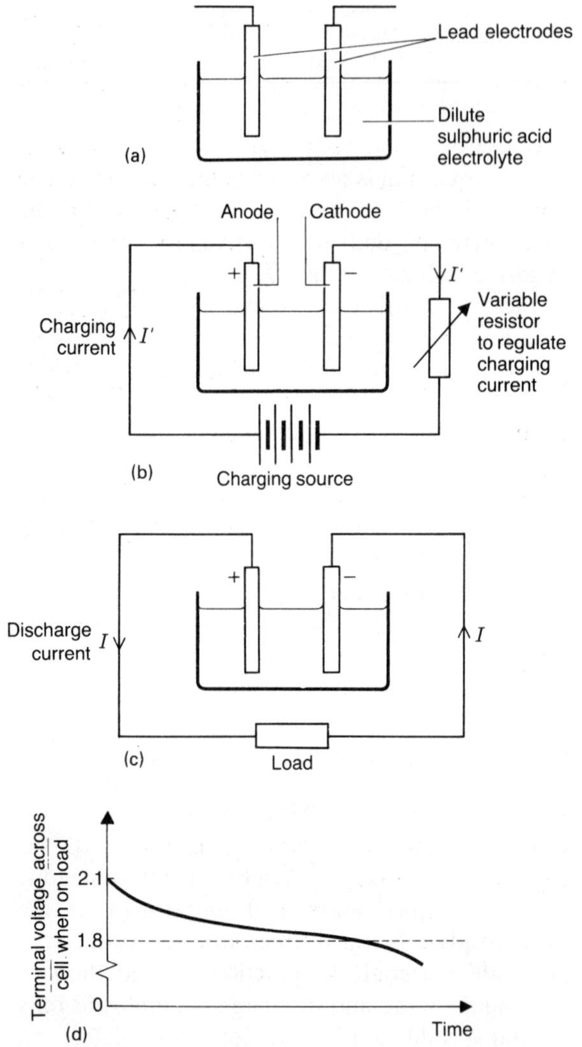

Figure 13.9
(a) A simple lead acid secondary cell (uncharged)
(b) Charging of lead acid cell to form active material on anode, and so create the e.m.f. of the cell
(c) Cell discharging, i.e. acting as a source of e.m.f.
(d) Variation of terminal voltage across cell when connected to a load, i.e. discharging

On a subsequent charge, current entering the positive plate from the electrolyte in the form of sulphate ions attacks the lead sulphate deposit and changes it to the active lead peroxide material. At the same time, hydrogen ions react with the lead sulphate coating on the cathode, converting this back into lead and also liberating hydrogen gas. At both plates, sulphuric acid is produced, thus replenishing the strength of the electrolyte.

Although commercial lead–acid cells are capable of supplying several amperes (and even several tens of amperes for a short time) care must be taken not to exceed the rated current value for a cell. If this were exceeded, buckling of the cell plates might occur and as a consequence the cell would be permanently damaged.

The capacity of a lead–acid accumulator is stated in terms of the quantity of electricity it can supply when uniformly discharged in a given time, assuming, of course, that it is initially fully charged and is finally not discharged beyond the point that it could not be fully charged again. The capacity is normally quoted in ampere-hours (A h). Thus a 40 A h cell could supply 2 A for 20 hours or 4 A for 10 hours (provided of course that the discharge current did not exceed the current rating). Strictly speaking the ampere-hour is not a recognized SI unit, but its use to specify cell capacity is very widespread. Note that

$$1\,\text{A h} = 1\,\text{A} \times (60 \times 60)\,\text{s} = 3600\,\text{As (ampere-seconds)} = 3600\,\text{C (coulombs)}.$$

A labelled diagram of a typical lead–acid accumulator is shown in Figure 13.10. The positive plate consists of the active lead peroxide material in the form of a paste pressed into a lead–antimony alloy (94% lead, 6% antimony) framework. The negative plate consists of a

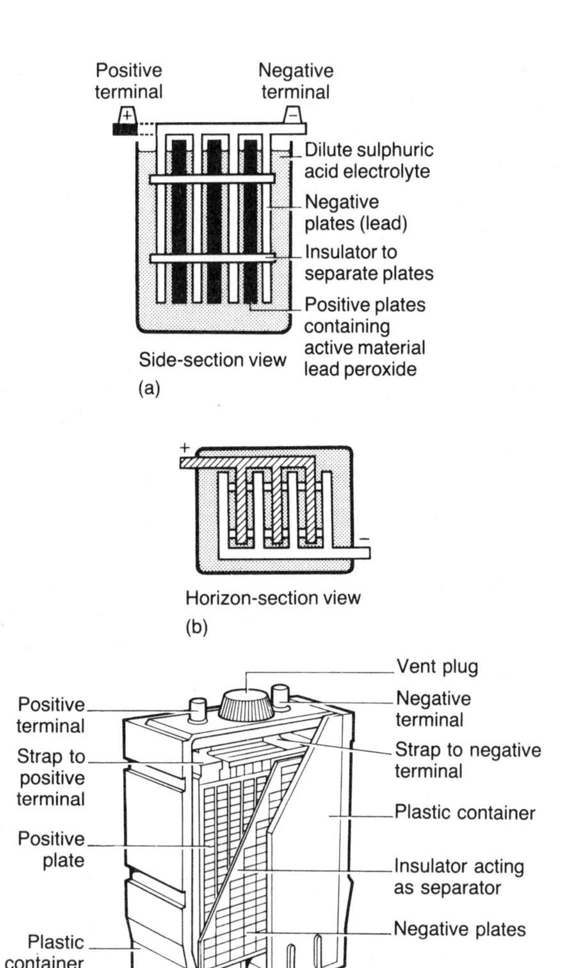

Figure 13.10 Lead–acid secondary cell accumulator

similar structure carrying lead in spongy form. To ensure a low internal resistance, a large plate surface area is needed, and to achieve this a number of positive and negative plates are interleaved, e.g. in Figure 13.10(a) there are four negative and three positive plates. The respective positive and negative plates are joined by a lead bar. To further reduce the internal resistance, the plates should be close together, but must be prevented from touching each other. Separators of plastic, glass, and 'treated' wood insulating materials, are used for this purpose. The electrolyte is dilute sulphuric acid which should have a relative density of about 1.21 when the cell is fully charged. The container itself is normally made of a hard durable plastic.

13.7 Effect of internal resistance on source terminal voltage

The voltage measured across the two terminals of a source of electrical energy, when these terminals are not connected to any external component or network, is equal to the e.m.f. of the source. Thus, in Figure 13.11(a), the voltage measured across the open terminals (assuming that the voltmeter takes negligible current), V_0 = e.m.f. of the source. Under these conditions, no current flows and we say that the source is on no-load, or open-circuited. The term 'load' is used in electrical circuits to denote the component or network connected across the source terminals (Figure 13.11(b)).

If our source is ideal, the voltage across its terminals remains constant and equal to the e.m.f. or no-load value of voltage when the source actually delivers current to a load con-

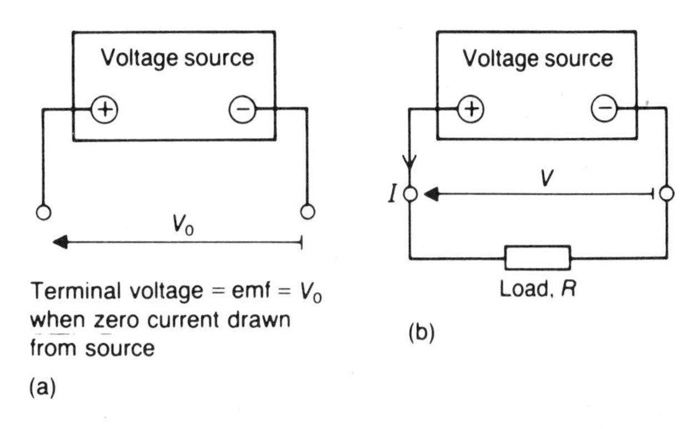

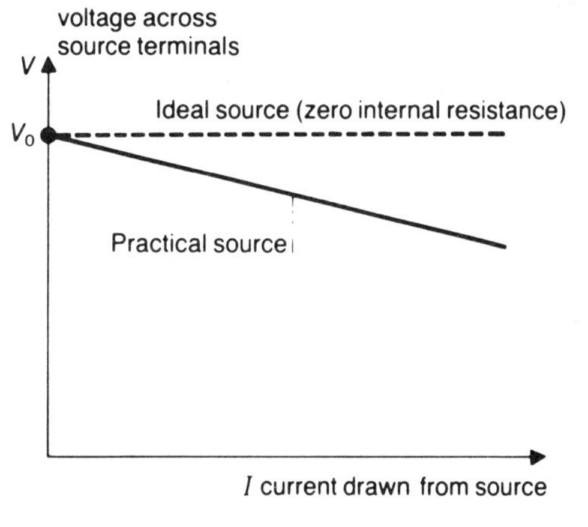

Figure 13.11

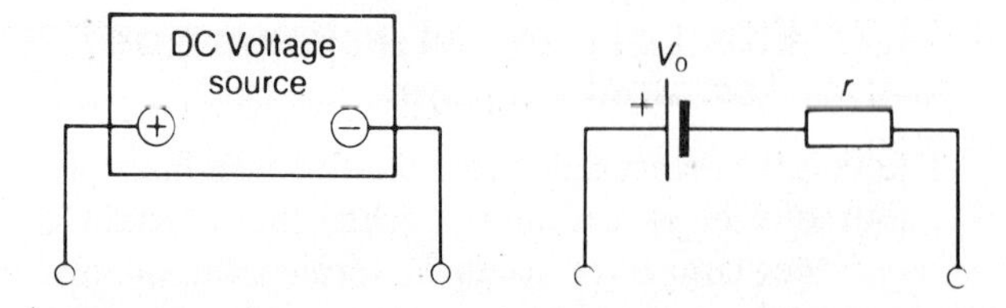

Figure 13.12 Equivalent circuit for a practical voltage source V_o *= e.m.f.,* r *= internal resistance*

nected across its terminals. However, for a practical source, the terminal voltage decreases as the load current increases. This effect is shown graphically in Figure 13.11(c).

Let us attempt to give an explanation of this decrease in terminal voltage with increasing load current by constructing a simple circuit model for a practical source, and then using this model to represent the electrical performance of the source in a circuit. A circuit model of a device must simulate accurately the behaviour of the device in an electrical circuit. In fact, one of the simplest examples of a circuit model is that of a linear resistor, defined by Ohm's Law as $R = V/I$, which simulates the behaviour of a conductor in a circuit.

A practical source will always exhibit some resistance to the flow of current and hence energy will be expended in driving current within the source itself as well as in the externally connected circuit components. This effect may be taken into account by representing the source between its two terminals by a model consisting of a voltage equal to the source e.m.f. of V_0 volts in series with a resistance r. The circuit model for a practical source is shown in Figure 13.12. The resistance r is known as the **internal resistance** of the source. It simulates the resistance to current flow and associated heat dissipation effects within the source itself, hence its name.

Next, consider the source connected across a load of resistance R. The equivalent circuit for this situation is drawn in Figure 13.13. On analysing the circuit using Ohm's law and the fact that the applied e.m.f. equals the sum of the voltages across the internal resistance r and the load resistance R, we have

$$V_0 = RI + rI = (R + r)I$$

so the circuit current,

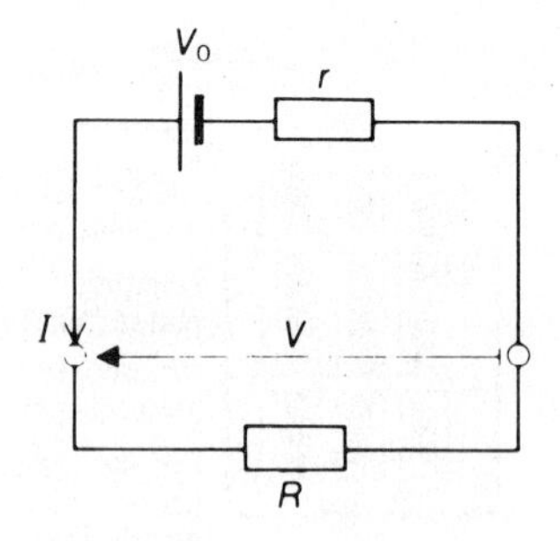

Figure 13.13 Source of e.m.f. V_o *and internal resistance* r *supplying current to load* R

$$I = \frac{V_0}{R + r}$$

and the voltage across the source terminals,

$$V = RI = \frac{R}{R + r} V_0$$

Thus if the load R is much greater than the internal resistance r,

$$\frac{R}{R + r} \approx 1 \text{ and so the terminal voltage } V \approx V_0$$

(the symbol $\approx$ means approximately equal to). However, if $R = r$

$$\frac{R}{R + r} = \frac{R}{2R} = \tfrac{1}{2} \text{ so } V = \tfrac{1}{2}V_0 \text{ (half the e.m.f.),}$$

whilst if $R = 0.1r$ (i.e. load resistance very much smaller than the internal resistance),

$$\text{then } \frac{R}{R + r} = \frac{0.1r}{0.1r + r} = \frac{0.1r}{1.1r} = \tfrac{1}{11}$$
$$\text{and } V = \tfrac{1}{11}V_0,$$

showing that the terminal voltage is less than 10% of the e.m.f. or no-load voltage.

Let us also consider the circuit from a power (or energy flow) point of view. One reason for doing this is that the formal definition of the e.m.f. of a source is often stated in terms of energy flow per coulomb of charge transported i.e. the e.m.f. of a source is equal to the energy supplied by the source to transport one coulomb of charge completely around the circuit to which the source is connected. Thus, using this definition, the total energy per second (i.e. the power) supplied by the source

$= (\text{e.m.f.}) \times (\text{charge per second}) = V_0 I.$

The power dissipated within the source (as heat) $= rI^2$, the power dissipated in the load resistor $= RI^2$, and as energy and power are conserved, power supplied by source = power dissipated in r + power dissipated in R

so $V_0 I = rI^2 + RI^2.$

Cancelling through the above equation by I gives

$$V_0 = rI + RI$$

from which the source terminal voltage

$$V = RI = V_0 - rI.$$

Thus, we obtain the identical results to our first analysis, exactly as we should expect: the terminal voltage of a source decreases as the load current increases.

Examples

1 The open-circuit voltage across the terminals of a dc voltage source was measured by a voltmeter of very high resistance and was recorded as 24.0 V. A 100 Ω resistive load was then connected across the terminals and the terminal voltage was observed to fall to 22.5 V.
Determine the e.m.f. and internal resistance of the source.

Solution

Let V_0 = source e.m.f. and r = its internal resistance. On open-circuit no current is drawn from the source and so terminal voltage = source e.m.f.,

i.e. $V_0 = 24.0\,\text{V}$ *Ans*

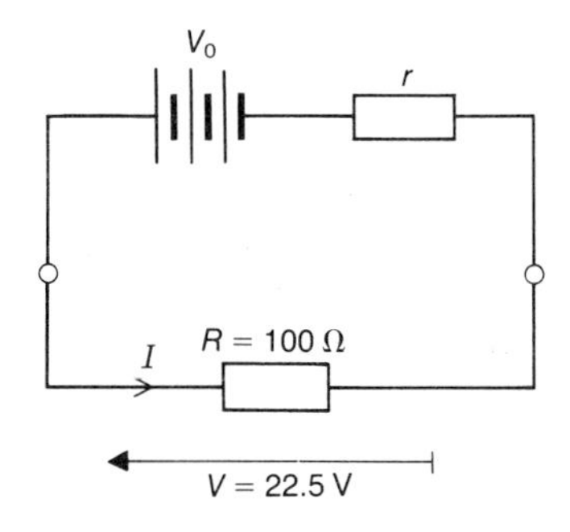

igure 13.14 For example 1

When the 100 Ω load is connected, see Figure 13.14, the terminal voltage, $V = 22.5$ and the load current,

$$I = \frac{V}{R} = \frac{22.5}{100} = 0.225\,\text{A}$$

The 'remainder' of the source e.m.f. is dropped across the internal resistance, so

$$rI = 24 - 22.5 = 1.5\,\text{V}$$

and $r = \dfrac{1.5}{I} = \dfrac{1.5}{0.225} = 6.67\,\Omega$ *Ans*

2 The following results were obtained for a terminal voltage versus load current test on a battery:

Terminal voltage (V)	3.2	3.1	3.0	2.9	2.8	2.65	2.45
Load current (mA)	0	24	48	74	99	147	196

Plot the graph of terminal voltage versus load current and determine the internal resistance of the battery.

Solution

The graph of terminal voltage V versus load current I is plotted in Figure 13.15 and a straight line drawn through the points. The graph shows that V falls as I increases, indicating the effect of internal resistance. The equation of the straight line is

$$V = V_0 - rI$$

where $V_0 = 3.2\,\text{V}$, the e.m.f. of the source and value of V when $I = 0$ and r = internal resistance.

Thus $r = \dfrac{V_0 - V}{I}$

and from the graph, we have $V = 2.45\,\text{V}$ when $I = 200\,\text{mA}$

so $r = \dfrac{3.2 - 2.45}{200 \times 10^{-3}} = \dfrac{0.75}{0.2} = 3.75\,\Omega$

Ans

3 Six cells, each of e.m.f. 2.0 V and internal resistance 0.02 Ω are connected in series and charged from a 30 V source of negligible internal resistance. The charging current is to be set at 4 A. Calculate the value of the series

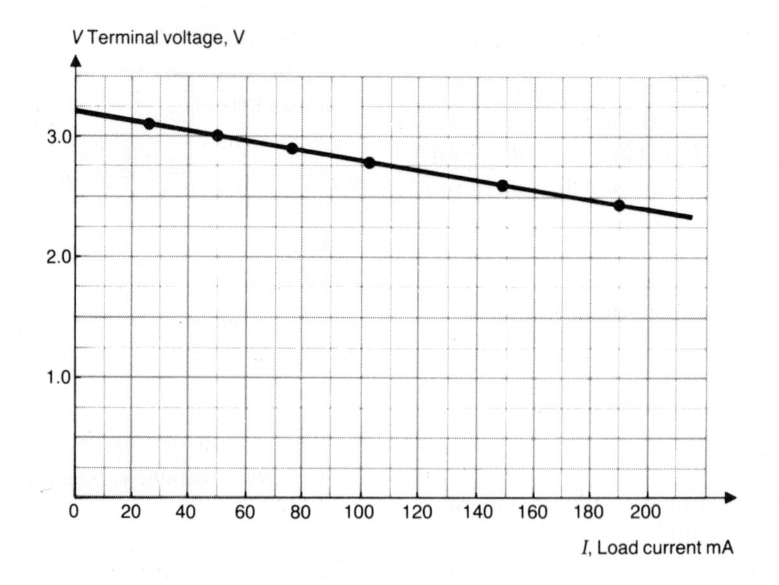

Figure 13.15 Plot of terminal voltage versus current, see example 2

resistance R that should be included in the circuit to achieve this.

Solution

The equivalent charging circuit is shown in Figure 13.16. The total e.m.f. of the six cells connected in series is $6 \times 2.0 = 12.0\,\text{V}$ and the total effective internal resistance is $6 \times 0.02 = 0.12\,\Omega$.

The resultant e.m.f. in the circuit is 30 V due to the charging source, less the 12 V due to the cells acting in opposition, so the charging current in the circuit can be found from:

$$30 - 12 = (I \times 0.12) + (I \times R)$$

setting $I = 4\,\text{A}$, the required current, we have

$$18 = (4 \times 0.12) + 4R$$

$$\text{so} \quad 4R = 18 - 0.48 = 17.52$$

$$R = \frac{17.52}{4} = 4.38\,\Omega \quad \textit{Ans}$$

Test 13

This test may be used as a basic self-assessment test to check your understanding of Chapter 13 on **Electrochemical effects and applications** and its learning objectives. Enter all answers in the answer block.

Qu. 1 Enter a tick (√) in the answer block if you consider the statement is correct; enter a cross (×) if you consider the statement is in any way incorrect.

(a) Metals are normally good conductors of electricity since they can provide electrons to carry the current, while in insulating materials all electrons are tightly bound to their atoms.

(b) Conduction in electrolytes is by free ions and is accompanied by chemical decomposition.

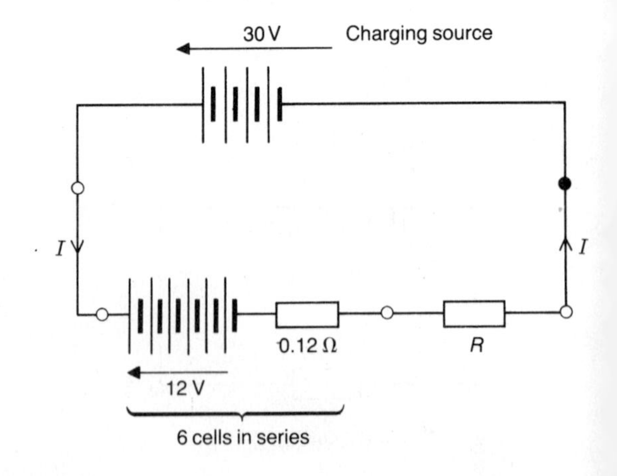

Figure 13.16 Charging circuit for example 3

Answer block:

Question no.		Ans	Qu. no.		Answer Qu. 2
1	(a)		2	(a)	
	(b)			(b)	
	(c)			(c)	
	(d)			(d)	
	(e)			(e)	
3	(a)			(f)	
	(b)				
	(c)				

(c) Distilled water is a very poor conductor of electricity since it contains very few free ions.
(d) Electroplating is the process of depositing a layer of a metal on an object connected as an electrode when a current flows in an electrolyte.
(e) A secondary cell is one in which the chemical action is reversible, it can supply current and may also be recharged.

Qu. 2 In Figure 13.17 electrode X is pure nickel, electrode Y is brass and the electrolytic solution is nickel sulphate ($NiSO_4$) containing nickel Ni^{++} ions and sulphate ions SO_4^{--}.

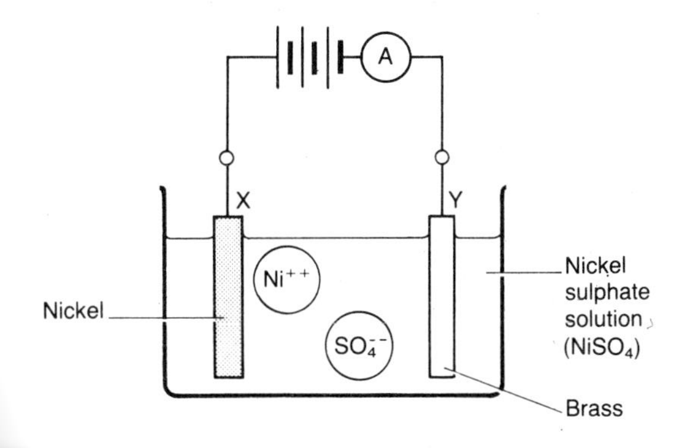

Figure 13.17 For Qu 2

Answer the following:
(a) Electrode X connected to the positive supply terminal is called the ...
(b) Electrode Y is called the ...
(c) Name the current carriers in the solution and state to which electrode they flow.
(d) State briefly what happens at electrode Y.
(e) State briefly what happens at electrode X.
(f) The chemical decomposition effect of a current flowing through an electrolyte is known as ...

Qu. 3 The following results were obtained on testing a voltage source:

Load current (A)	0	1	2	3	4	5
Terminal voltage (V)	10.0	9.7	9.1	8.8	8.3	8.1

Plot a graph of terminal voltage versus load current and determine:
(a) the source e.m.f.;
(b) the terminal voltage when the supply delivers 3.5 A;
(c) the internal resistance of the source.

Problems 13

1 (a) Give a brief explanation as to why copper is an excellent conductor of electricity

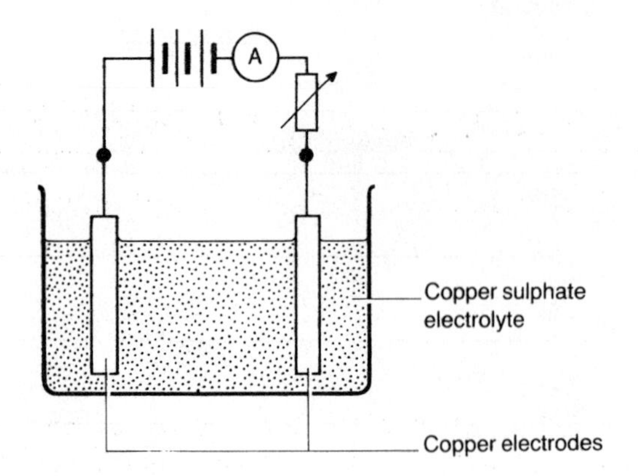

Figure 13.18 For problem 3

while ceramic and most plastic materials do not conduct electricity.

(b) Describe briefly why electrolytes conduct electricity while distilled water is a very poor conductor.

2 Explain with the aid of a diagram the process of electroplating.

3 Explain the meaning of the terms electrolytes, electrolysis and electrodeposition.
Describe the action taking place in the circuit of Figure 13.18 where both electrodes are copper and the electrolyte is copper sulphate. It is known that 96 500 C of charge can liberate 31.77 g of copper in electrolysis. Using this information determine the current that should flow in Figure 13.18 to cause 5 g of copper to be deposited on the cathode per hour.

4 Describe the essential differences between primary and secondary cells.
Draw a labelled diagram of a lead–acid accumulator and describe how the accumulator may be charged.
Sketch curves showing the variation of terminal voltage with time when supplying current (discharge) and also during charge.

5 Describe an experiment to obtain data relating terminal voltage of a source to load current supplied.

Load current (A)	0	0.5	1.0	1.5	2.0	2.5
Terminal voltage (V)	5	4.8	4.5	4.4	4.1	4.0

Plot the graph of terminal voltage versus load current for the above results and determine the e.m.f. and internal resistance of the voltage source.

6 Define the e.m.f. and internal resistance of a voltage source.
A battery is connected in series with an ammeter and a variable resistance. The following results are recorded:

Ammeter reading (mA)	75	50	25	15
Resistance (Ω)	18	28	58	98

Draw a circuit diagram representing the battery by its e.m.f. V_0 and internal resistance r and hence show that the circuit current is given by

$$I = \frac{V_0}{R + r}$$

Use the above results to deduce the value of V_0 and r for the battery.

7 A battery of e.m.f. 12 V and internal resistance 20 Ω supplies current to a 100 Ω load. Calculate the voltage across the load when measured by an ideal voltmeter and when measured by a voltmeter of resistance 400 Ω.

8 Three similar lead–acid cells which have e.m.f.s of 2.1 V and internal resistances of 0.06 Ω are connected (a) in series; (b) in parallel, to supply a load of 2 Ω.
Calculate the load current in each case.

9 Six 1.2 V nickel–iron cells, each of 0.2 Ω internal resistance are connected in series and charged from a 20 V source. A variable resistance R is connected in series in the circuit to control the value of the charging current. Calculate the value of R if the charging current is to be set at (a) 1 A; (b) 5 A

14 Electromagnetic effects and applications

General learning objectives: to establish an understanding of electromagnetic effects and to explain practical applications of these.

14.1 Introduction to magnetism: magnets and magnetic effects of electrical current

Magnetism is the branch of physics concerned with magnets, magnetic materials and magnetic effects. Electromagnetics or electromagnetism combines both magnetism and the magnetic effects produced by electric currents and also the important ideas that a changing magnetic field can produce an e.m.f. In this chapter we consider the magnetic and electrical effects associated with magnetic fields and their applications, but first we introduce some basic ideas relating to magnets and magnetic fields produced by magnets and conductors carrying dc current.

The metals of iron, cobalt and nickel, and certain alloys, exhibit very strong magnetic properties and are classed as ferromagnetic materials. Other metals such as copper and aluminium exhibit no such effects.

A magnet consists of ferromagnetic material which has been magnetized to have a permanent magnetic effect and may be made in a variety of configurations. Two common examples, the bar and horseshoe magnets are shown in Figure 14.1(a). A magnet appears to have its magnetism concentrated at two main points, which are called poles: the north (N) pole and the south (S) pole. A magnet exerts force on other ferromagnetic materials, it will attract such materials towards its poles.

If a bar magnet is suspended so that it can swing freely in the horizontal plane, then the N pole will eventually line up so that it points towards

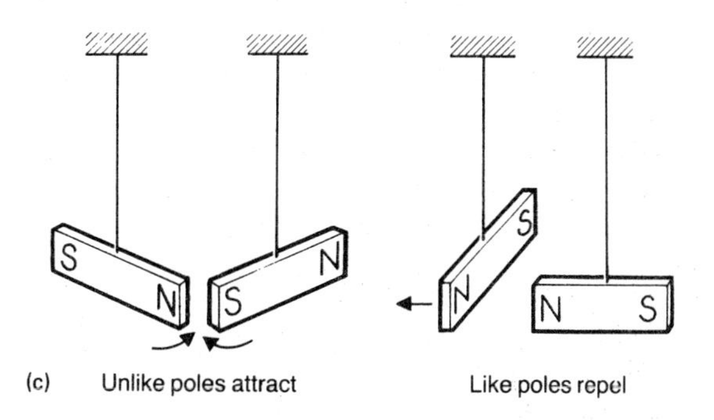

Figure 14.1
(a) Examples of magnets
(b) A suspended magnet points to magnetic north.
(c) Forces between magnets

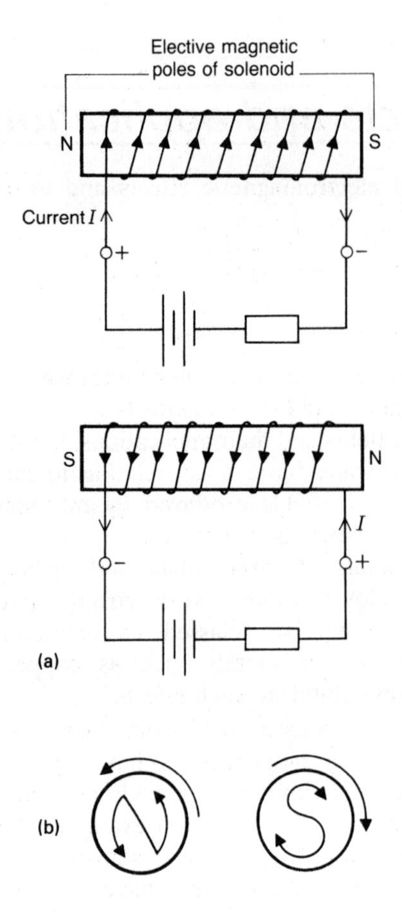

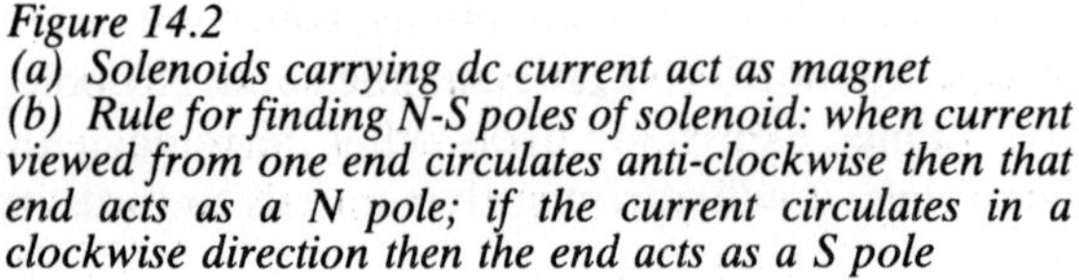

Figure 14.2
(a) Solenoids carrying dc current act as magnet
(b) Rule for finding N-S poles of solenoid: when current viewed from one end circulates anti-clockwise then that end acts as a N pole; if the current circulates in a clockwise direction then the end acts as a S pole

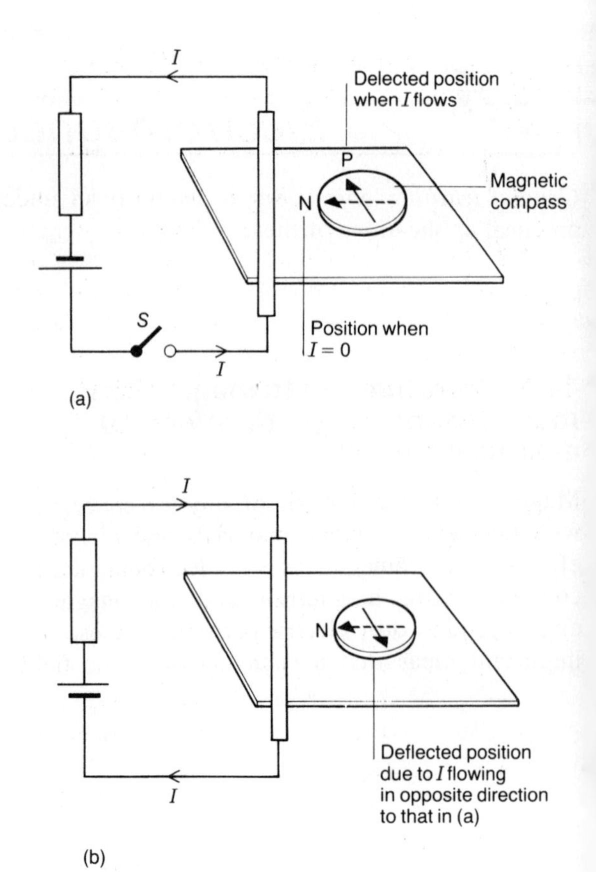

Figure 14.3 Demonstration that an electric curren[t] produces a magnetic effect

magnetic north and the magnet itself is in line with the earth's magnetic field, see Figure 14.1(b) – this fact forms the principle of the magnetic compass and has been used for centuries in navigation. If the pole of a magnet is brought close to the 'unlike' pole of another magnet, i.e. N to S or S to N, then a force of attraction between the two poles is observed; if the pole are the same, i.e. N to N or S to S then a repulsive force is observed, see Figure 14.1(c).

Current-carrying conductors also produce magnetic effects. For example, when the solenoidal coils shown in Figure 14.2(a) are energized to carry dc current, they produce identical magnetic effects as bar magnets. Magnetic poles are produced at each end of the solenoid. A 'rul[e' for determining the respective N and S poles [a]t the ends of a solenoid with respect to th[e] direction of the current flow in the windings [is] given in Figure 14.2(b).

A simple illustration of the magnetic effect of [a] conductor carrying current is also shown in Figur[e] 14.3. When the switch S is open, no current flo[ws] in the circuit and the magnetic compass needl[e] points in the direction of magnetic north N. Whe[n] the switch is closed the circuit is complete, curre[nt] flows in the conductor and the compass needle [is] deflected. If the direction of the current [is] reversed the deflection of the compass needle is [i]n the opposite direction.

14.2 Applications of magnetic force effects: electromagnets and magnetic relays

If a solenoid is wound on a bar or rod of ferromagnetic material and a dc current is established the material becomes magnetized and the bar or rod behaves as a magnet. When the current is switched off the magnetic properties are very much reduced and in the case of soft iron it will no longer act as a magnet. This principle is applied in the electromagnet which consists in its simplest form of a soft-iron rod on which is wound a solenoid of insulating wire, as shown in Figure 14.4(a).

When current flows in the windings of an electromagnetic, powerful magnetic poles are formed which are capable of attracting and lifting pieces of iron and steel. When the current is switched off, the electromagnet loses most of its magnetism and hence gravity will cause the release of any material that it had previously attracted. One important application of electromagnets, see Figure 14.4(b), is as a 'magnetic' crane in scrap-yards.

The attractive force on magnetic materials produced by solenoids or windings acting as magnets when energized by electric current also finds important application in magnetic relays, two examples of which are shown in Figure 14.5. In (a), when a relatively small dc current I flows through the relay windings, the soft-iron core P is attracted downwards by the magnet so formed, closing the break in the main circuit by placing a conductor between terminals A and B. This conductor completes the main circuit and allows the main circuit current to flow. On opening switch S in the relay-control circuit, I falls to zero with the result that the relay windings are no longer energized and lose their magnetic effect. The return spring forces the conductor bar upwards, breaking and switching off the main circuit. Thus a small current in the control circuit can control the switching and operation of the main circuit.

A second example is shown in the electric bell circuit of Figure 14.5(b). The iron disc attached to the bell striker is attracted towards the electromagnet whenever current flows in its windings. However, this process breaks the circuit resulting in removal of the magnetic effect of the electromagnet and a spring returns the striker to its original position. Current can now flow again and the cycle of events is repeated with the result of ringing the bell at each forward stroke.

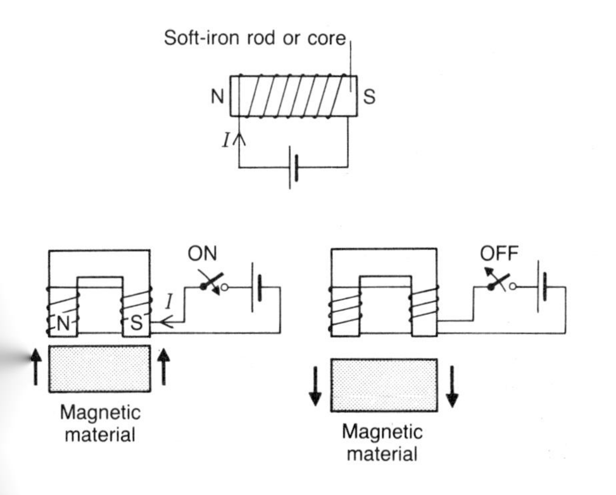

Figure 14.4 Electromagnets

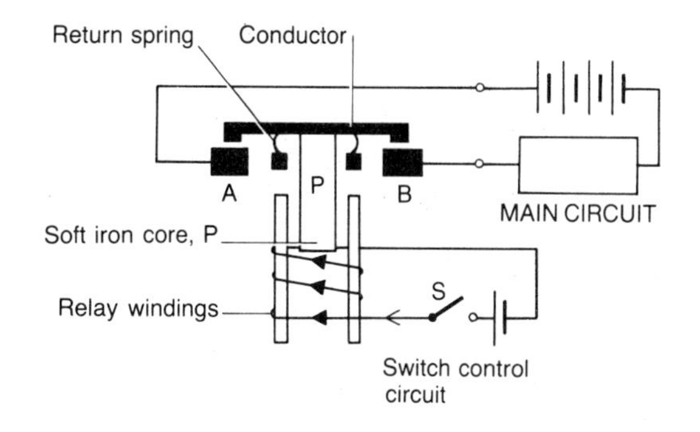

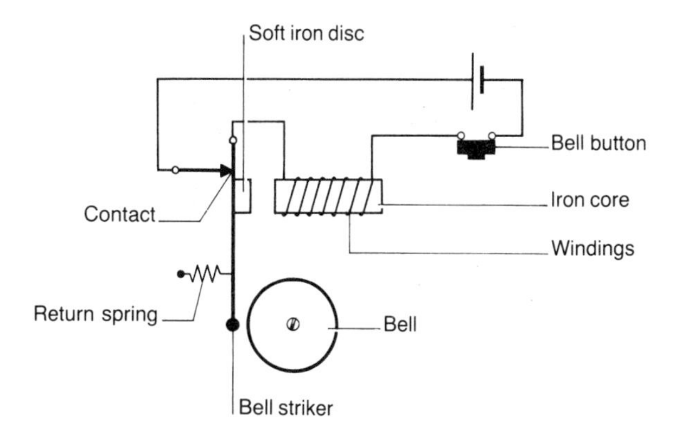

Figure 14.5 Magnetic relays: examples of application

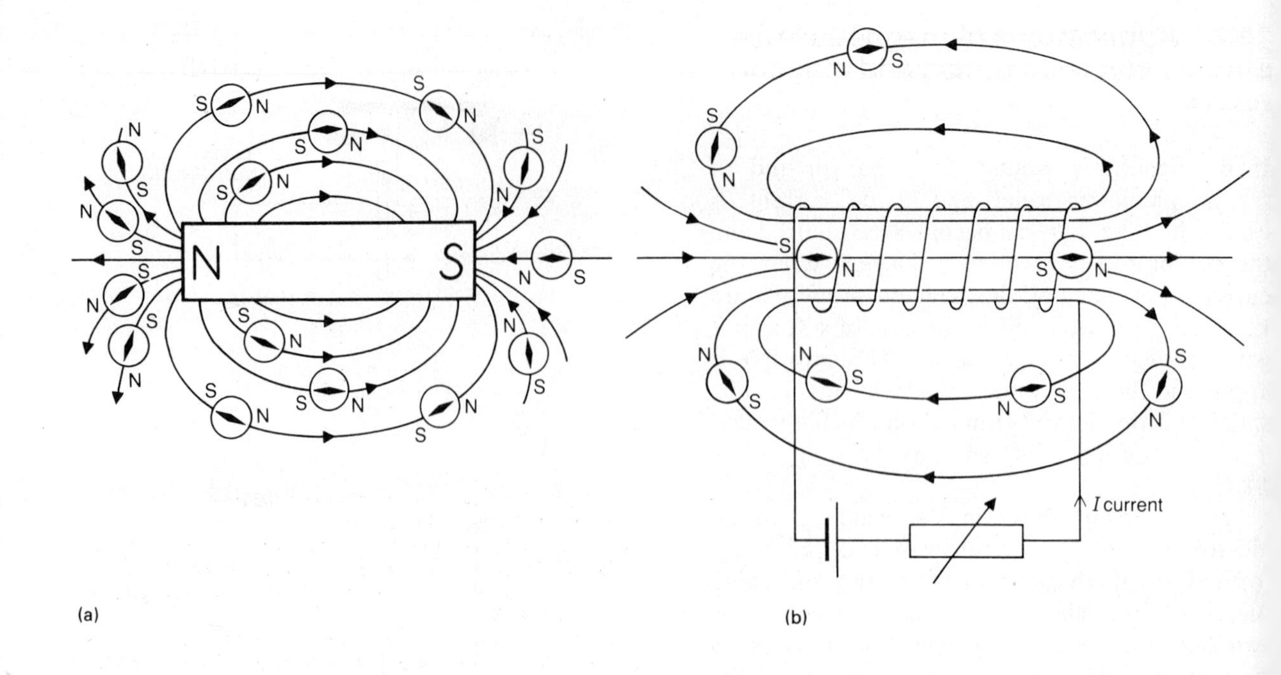

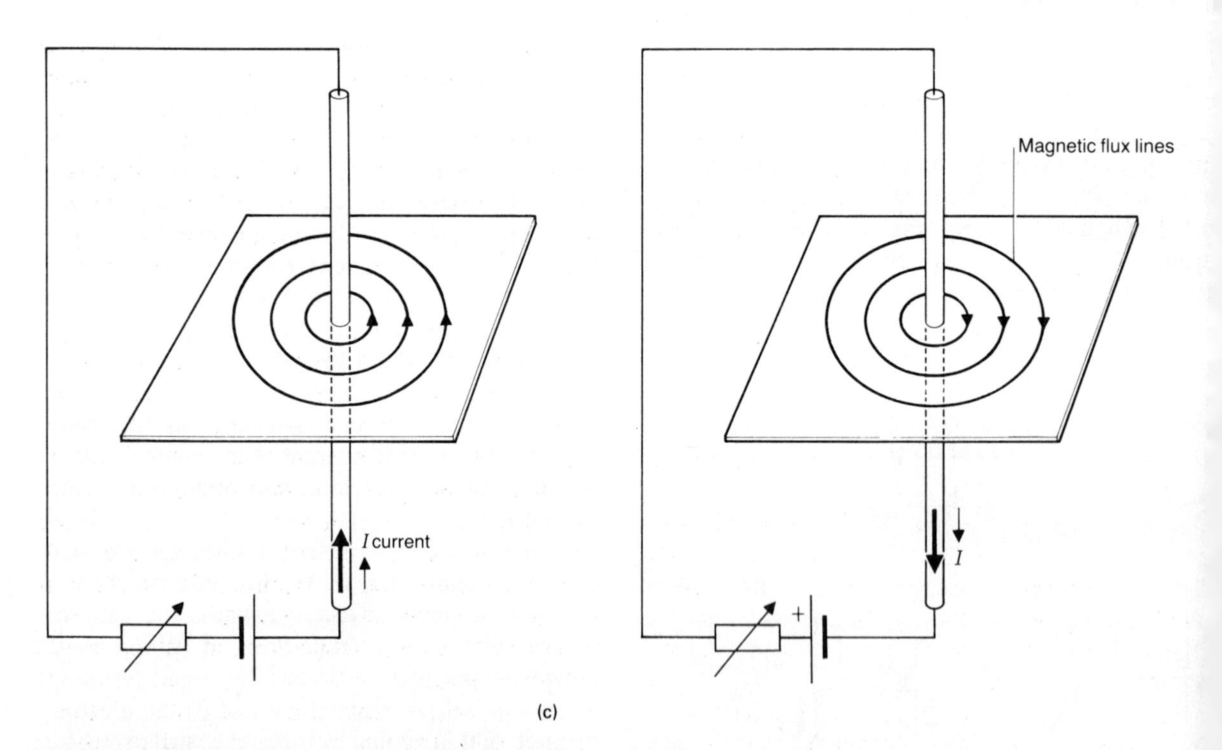

Figure 14.6 Magnetic field pattern for bar magnet, solenoid and single current-carrying conductors
(a) Magnetic pattern around a bar magnet
(b) Magnetic field pattern around a solenoid
(c) Magnetic field pattern in a plane perpendicular to a single current-carrying conductor

14.3 Magnetic fields and the concept of magnetic flux density

A magnet or a current-carrying conductor produces in its surrounding space forces on other magnets, ferromagnetic materials and current-carrying conductors brought into this region. To describe these effects, we say a magnet or a conductor carrying current produces a magnetic field and we represent the 'pattern' of a magnetic field by means of imaginary lines known as lines of magnetic flux. These lines are lines of force and the direction of a line at any point is the direction in which the N pole of a tiny magnetic compass would point if placed at the point. It is important to emphasize that when using the concept of lines of magnetic flux which connect points of equal magnetic flux intensity that the lines should never touch or cross each other.

Figure 14.6 shows the magnetic field patterns produced by a bar magnet and a solenoid energized by dc current. The lines of magnetic flux may be plotted by using a small magnetic compass. Consider, for example, the plotting of a magnetic field due to a bar magnet. Place the compass at a point close to its north pole and record the direction in which the compass needle points by drawing an arrow; repeat for a series of other points at suitable intervals and join up the arrows – the tip of one to the base of the next. In this way a continuous path can be traced. By repeating the procedure, other field lines may be drawn in and a complete set surrounding the magnet obtained. The set of these magnetic field or flux lines is known as the magnetic field pattern. The direction of a field or flux line is taken as the direction in which the north pole of the magnetic compass would point if placed there. It represents the direction of the force on a N pole at the point in question.

The magnetic field pattern due to a solenoid could be obtained in a similar manner. In fact the field lines could be plotted within the solenoid itself showing that they form closed loops as shown in Figure 14.6(b). In (c) is shown the magnetic field pattern produced by a single straight conductor when carrying current. The flux lines in a plane perpendicular to the conductor form concentric circles and the direction of the flux depends on the direction of the current flow.

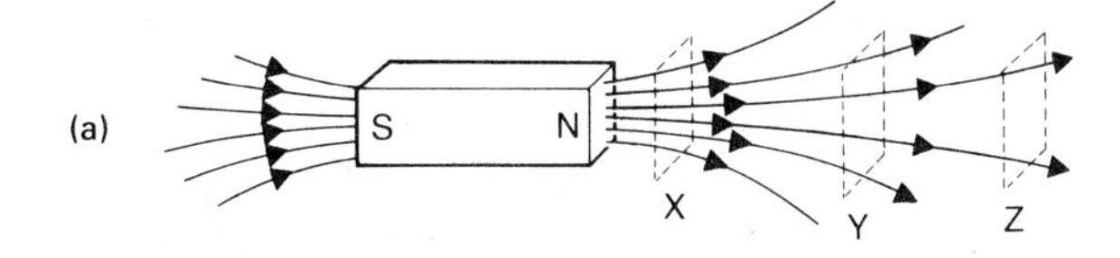

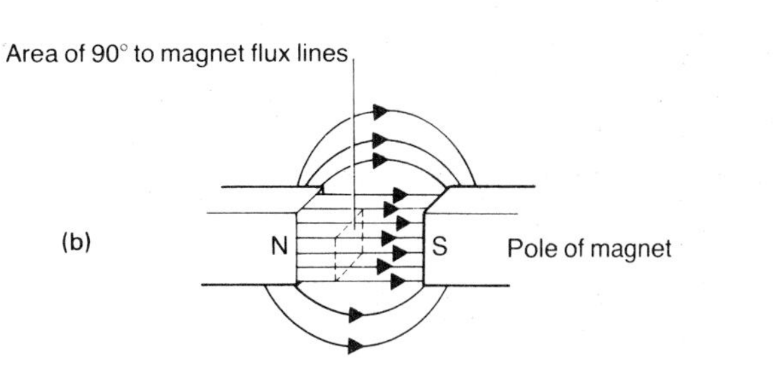

Figure 14.7 The concept of magnetic flux density B*:*
B = *magnetic flux/area* = Φ/A
(a) Magnetic flux lines in the vicinity of the axis of a bar magnet. Magnetic flux density decreases as we move away from the poles of the magnet, e.g. it decreases in direction XYZ
(b) Magnetic flux density B can be identified qualitively as directly proportional to the number of field or flux lines flowing at 90° through a unit area

In Figure 14.7(a) we have redrawn the magnetic field pattern of a bar magnet showing in more detail the flux lines along and in the vicinity of the axis of the magnet. It will be seen that the magnetic field pattern is such that the flux lines are widely separated when they are a long way from, and roughly parallel to the axis (e.g. region Z), but near to the poles of the magnet (e.g. region X) they crowd closely together. Exactly similar regions will also exist for the case of the solenoid coil of Figure 14.6(b). It is readily found, by placing a small piece of steel such as a pin in regions X and Z, that, while at Z the pin may well not move at all, at X it is pulled powerfully toward the magnet. At an intermediate point, such as Y, its motion will be, initially, more sluggish. Thus we may deduce that the more the flux lines are crowded together, the stronger is the magnetic field. This leads to the concept of magnetic flux density in quantifying the strength of a magnetic field.

The magnetic flux density at a given point in a magnetic field can be identified as directly proportional to the number of flux lines passing at right angles through a small area placed at the position of the point, as shown for example in

Figure 14.7(b). The greater the number of lines passing through this area, the stronger the field.

The magnetic flux density may be taken as a measure of the magnetic field strength or intensity* at a given point in a magnetic field. It is denoted by the symbol B and may be defined as

$$\text{magnetic flux density } B = \frac{\Phi}{A}$$

where Φ = magnetic flux passing through a small area A positioned at 90° to the flux lines.

The magnetic flux density is a vector quantity in that it has a direction as well as a numerical value. The direction of B is in the direction of the flux lines at the given point.

The SI unit of magnetic flux is the weber (Wb). The SI unit of magnetic flux density is the tesla (T). Note that since magnetic flux density equals flux divided by area, we have

1 tesla = 1 weber per square metre,

i.e. $1\,\text{T} = 1\,\text{Wb/m}^2$.

14.4 The force on a current-carrying conductor in a magnetic field

A conductor carrying current experiences a force when placed in a magnetic field. The magnitude of the force is proportional to the current and the length of the conductor perpendicular to the lines of magnetic flux and increases directly with the field strength as measured by the magnetic flux density B. When the magnetic lines of flux are at right-angles to the direction of current flow, the force is given by

$$F = BIL \text{ newtons}$$

where B = magnetic flux density of the field, units teslas (T)

I = current in amperes (A)

L = length of conductor in metres (m)

*The term magnetic field strength, denoted by the symbol H, is also used to define the intensity of a magnetic field. It has the units of amperes per metre (A/m). B and H are related by the formula $B = \mu H$ where μ is known as the permeability of the medium in which the magnetic flux passes. μ can normally be considered as a constant for a given medium. To avoid confusion in the present text, we will consider B only. B is the fundamental quantity when we are considering forces in magnetic fields.

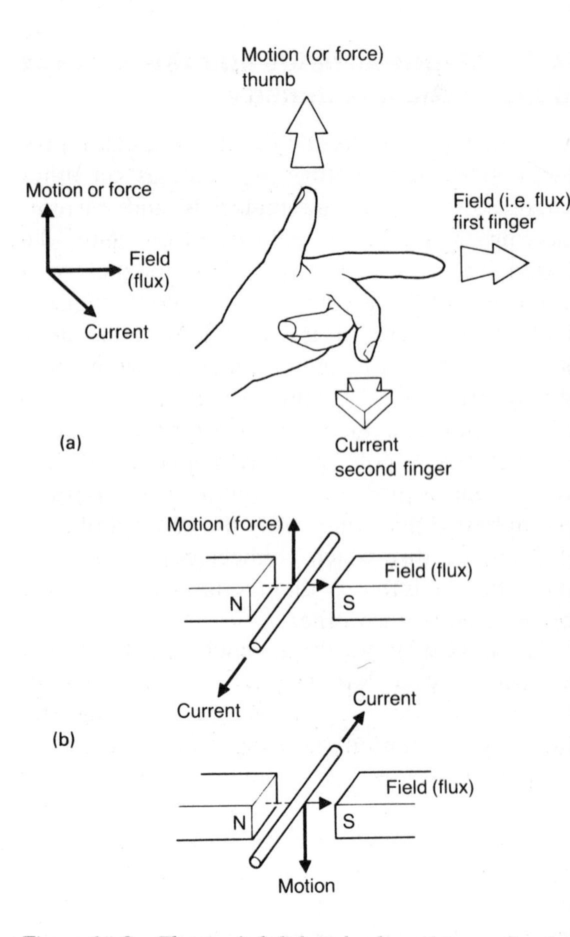

Figure 14.8 Fleming's left-hand rule and its application
(a) Fleming's left-hand rule to determine direction of force or motion of a current-carrying conductor in a magnetic field
(b) Examples showing the direction of the force on a conductor

The direction of the force is at right angles to both the directions of the current and the magnetic flux lines.

A useful aid for determining the direction of the force on a current carrying conductor is given by Fleming's left-hand rule, illustrated in Figure 14.8(a). This shows the left hand, with the thumb and first and second fingers so positioned that they are mutually at right angles. If the first finger is then pointed along the direction of the magnetic flux or field lines and the second finger in the direction of the current flow, then the thumb gives the direction of the force of the conductor, or if the conductor were free to move the direction of its motion. Fleming's left-hand

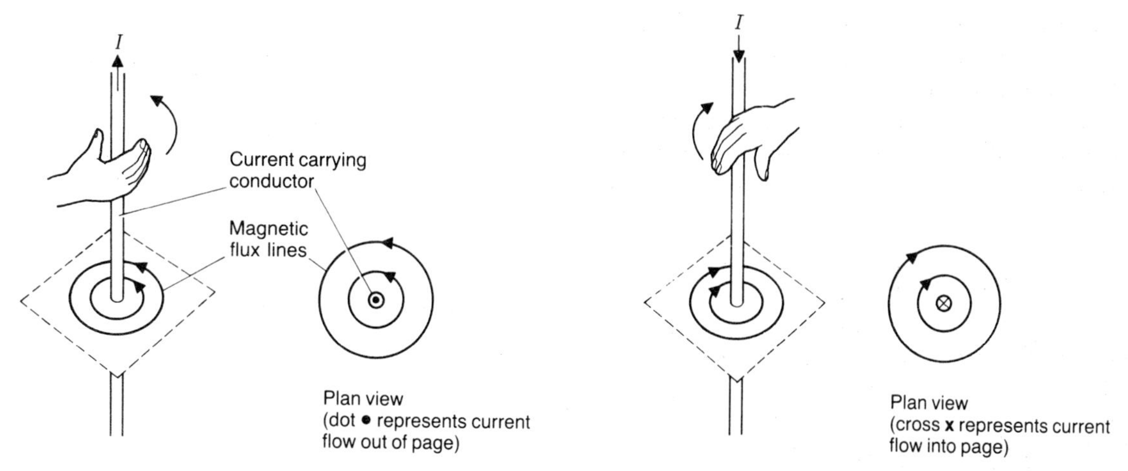

(a) Circular magnetic field patterns and the right-hand rule for determining the direction of the fluxlines for a straight current-carrying conductor

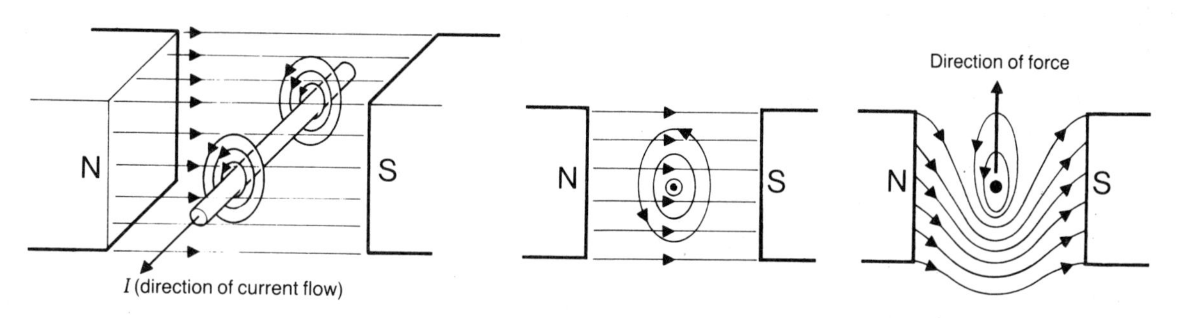

(i) A current-carrying conductor in the magnetic field between the poles of a bar magnet

(ii) Side view showing flux lines of conductor and magnet

(iii) Side view showing combined resultant magnetic field pattern and direction of force on conductor

(b) Magnetic field patterns when a current-carrying conductor is placed at right-angles to the field between the poles of a magnet

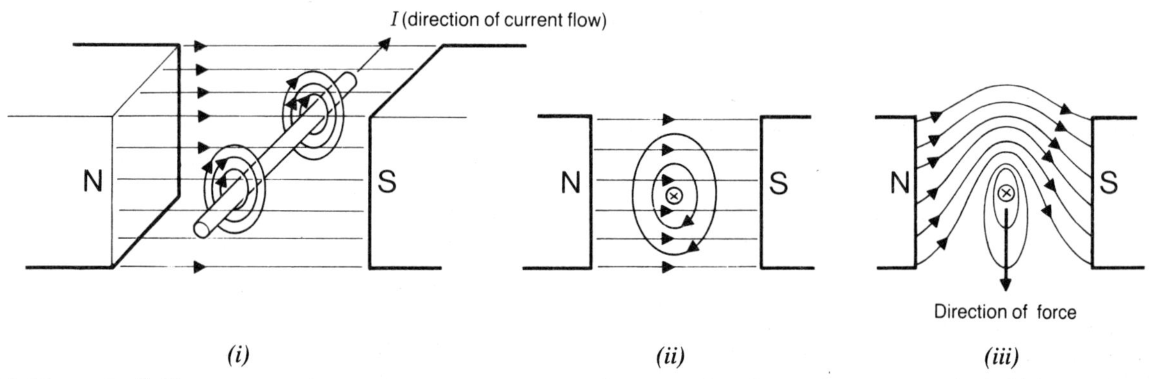

(i) *(ii)* *(iii)*

(c) Magnetic field patterns when a current-carrying conductor is placed at right-angles to the field between the poles of a magnet; current direction is opposite to that of (b)

Figure 14.9 Physical insight of Fleming's left-hand rule for determing the direction of force for a current-carrying conductor in a magnetic field

rule is applied in Figure 14.8(b). Check to see that you agree with the direction of the forces shown.

Let us now look at the plausibility of Fleming's left-hand rule from a physical viewpoint. Consider Figure 14.9(a) which shows the circular magnetic field pattern of a straight current-carrying conductor. A convenient way to remember the direction of the circular flux lines is the *right-hand rule*; if one 'holds' the current-carrying conductor with the right hand with the thumb pointing in the direction of the current, then the fingers will indicate the direction of the flux. Next, with the field pattern of the current-carrying conductor in mind, let us consider the effect of placing the conductor at right angles to a magnetic field, as shown for example in (b) and (c). In (b) (side view (ii)) the lines of flux produced by the current-carrying conductor and by the magnet below the conductor are in the same direction (left to right), while those above the conductors are in opposing directions. This results in a strong magnetic field below and a relatively weak field above the conductor. The combined field is shown in (b), (iii). Now, if we can think of the flux lines as like elastic cords distorted from their natural position and to which they will tend to return, the effect of the resultant field is to move the conductor vertically upwards. Check that this is in agreement with Fleming's left-hand rule, (c) shows that when the current direction is reversed the resultant field (see (iii)) is strongest above the conductor and thus will tend to move the conductor vertically downwards.

If the directions of the magnetic flux density B and current flow I are not at right angles, the force direction may still be found using Fleming's left-hand rule but the force magnitude is modified and is given in general as

$$F = BIL \sin\theta$$

where θ = angle between B and I directions.

Thus, with reference to Figure 14.10,

(a) $F = BIL \sin 90° = BIL$, as $\sin 90° = 1$
i.e. $\theta = 90°$ and B and I are therefore at right angles

(b) $F = BIL \sin\theta$
and if $\theta = 30°$, $\sin\theta = \sin 30° = 0.5$
so $F = 0.5\ BIL$

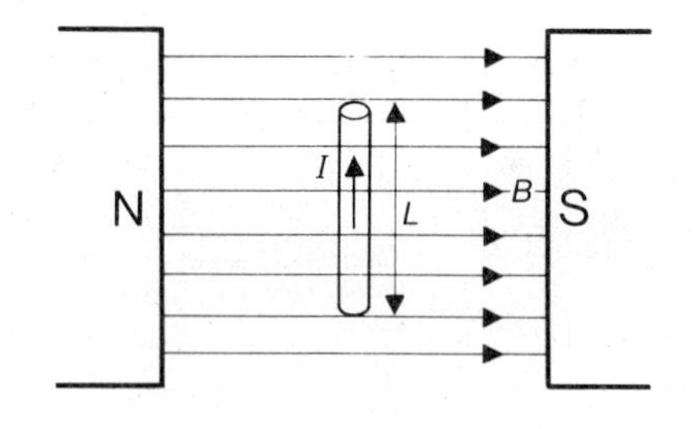

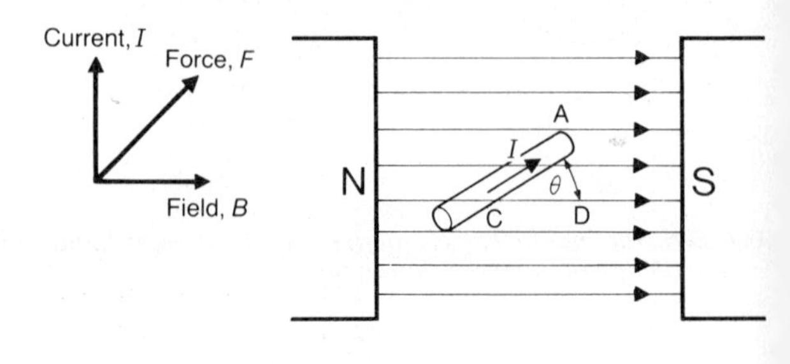

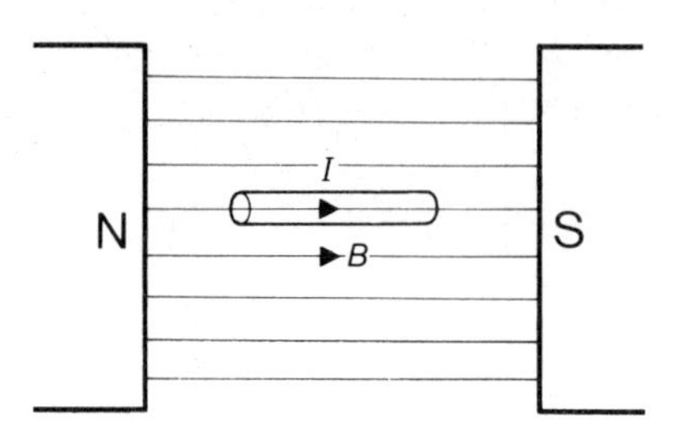

Figure 14.10 Force on a current-carrying conductor is given in general by F = BIL *sin* θ

(c) Here $\theta = 0°$ and as $\sin 0° = 0$, the force
$F = BIL \sin 0° = 0$

i.e. if the directions of the magnetic flux lines and current flow are parallel, the force is zero.

This is an important result. No force is experienced by a conductor carrying current when the directions of current flow and magnetic flux lines are in the same or opposing directions. The force on a current-carrying conductor is at a maximum when the current and the magnetic flux density are at right angles. The direction of the force in all cases is at right angles to both B and I and is given by Fleming's left-hand rule.

A simple experiment set up to demonstrate both the magnitude and direction of the forces experienced by a conductor carrying current in a magnetic field is shown in Figure 14.11. The conductor consists of a fine length of wire clamped between two fixed supports. The magnetic field is produced by either (a) a bar magnet, or (b) a solenoid. The following results may be observed:

1 When the current I flows in the wire, the wire moves downwards in the figures shown. If the bar magnet is turned through 180° or the direction of the current in the solenoid is reversed (i.e. the magnetic flux direction is reversed), the wire moves upwards.
2 The direction of the movement of the conductor is also reversed if the direction of the current in the conductor is reversed.
3 The displacement (and therefore the force) of the conductor in the vertical plane is increased when the current is increased and when the magnetic flux density is increased. The former may be demonstrated by varying the current by means of the variable resistor R, the latter by bringing the bar magnet or solenoid closer to the wire, which increases the magnetic flux density.
4 If the magnet or solenoid is turned so the magnetic flux lines no longer 'cut' the wire at right angles, the displacement of the wire is reduced and is zero when the magnetic flux is parallel to the wire.

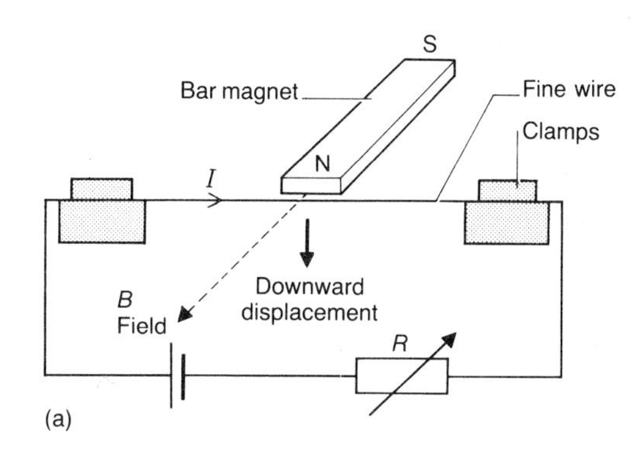

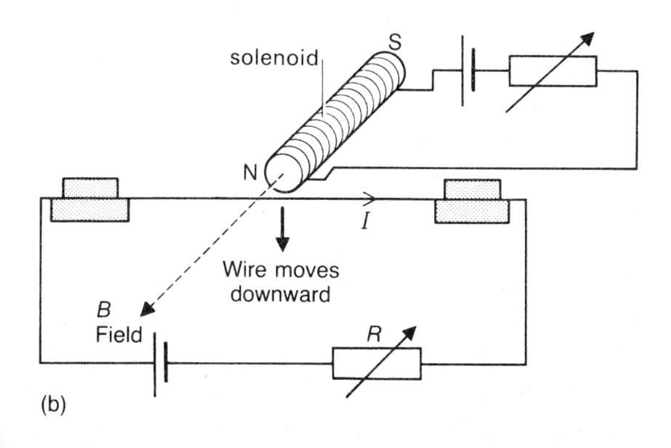

Figure 14.11 Demonstration of the force produced on a conductor-carrying current in a magnetic field

Examples

1 A straight wire of length 120 mm carries a current of 5 A and is situated in a magnetic field of flux density $B = 0.25$ T. The wire and direction of magnetic flux are at 90°. Calculate the force on the wire.
If the wire is moved to make an angle of 45° to B, calculate the new value of force on the wire.

Solution

The force on the wire when the current and magnetic fields are at 90° is

$$F = BIL$$

On substituting, $B = 0.25$ T, $I = 5$ A and $L = 120$ mm $= 0.12$ m, we have

$$F = 0.25 \times 5 \times 0.12 = 0.15 \text{ N} \quad \textit{Ans}$$

When B makes an angle of $\theta = 45°$ with I we have

$$\begin{aligned} F &= BIL \sin\theta \\ &= 0.25 \times 5 \times 0.12 \times \sin 45° \\ &= 0.15 \sin 45° = 0.15 \times 0.7071 \\ &= 0.106 \text{ N} \quad \textit{Ans} \end{aligned}$$

as sin 45° = 0.7071.

2 The conductors shown in Figure 14.12 are each 50 mm in length and carry a current of $I = 8$ A and are situated in a uniform magnetic field of $B = 4$ T. Calculate the force on the conductor in each case and state its direction.

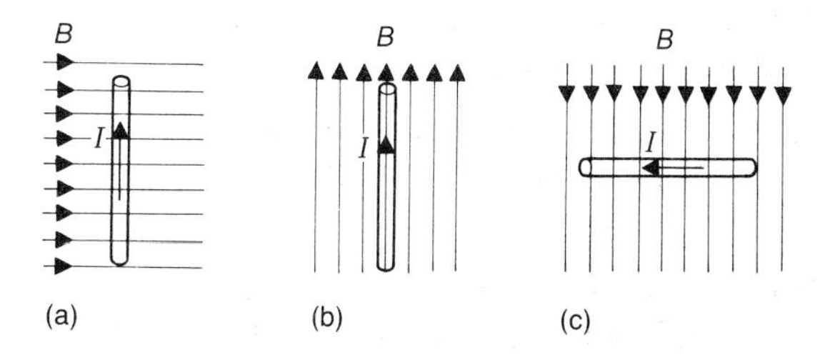

Figure 14.12 For example 2

Solution

(a) Since B and I are at right angles, the force on the conductor

$$F = BIL = 4 \times 8 \times 0.05,$$
$$\text{as } L = 50\,\text{mm} = 0.05\,\text{m}$$

so $F = 1.6\,\text{N}$ *Ans*

Using Fleming's left-hand rule we find that the direction of F is at right angles to both B and I and 'into' the page.

(b) Since in this case the directions of B and I are parallel, there is zero force on the conductor,

$F = 0$ *Ans*

(c) $F = BIL = 4 \times 8 \times 0.05 = 1.6\,\text{N}$ *Ans*

but in this case the direction of F, found by applying Fleming's left-hand rule, is 'out' of the page and at right angles to both B and I.

14.5 Application of force on a current-carrying conductor to current measurement: moving-coil meters

The force on a current-carrying conductor placed in a magnetic field is directly proportional to the

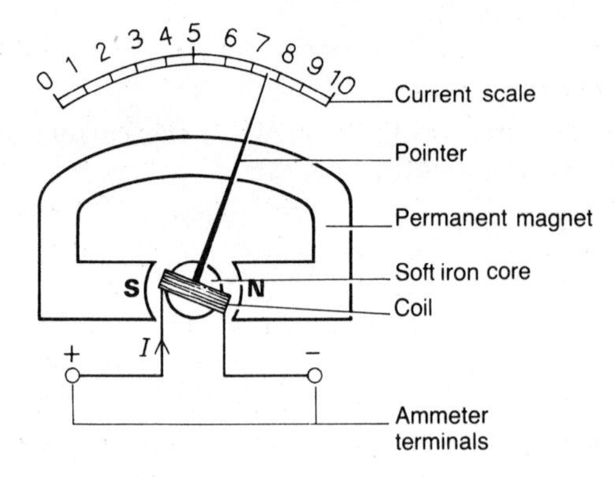

(a) Simplified diagram of a moving meter

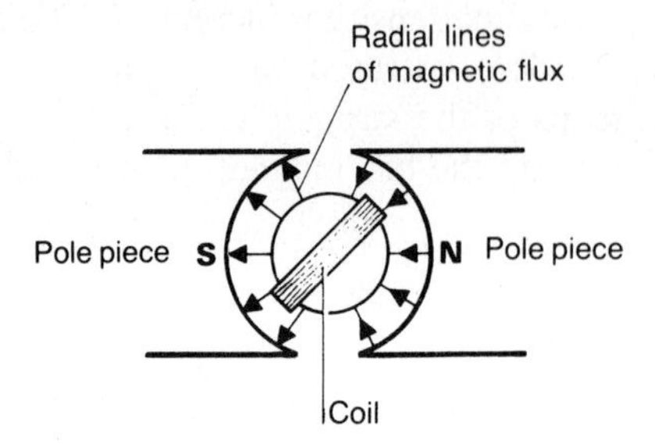

(b) Radial magnetic field pattern

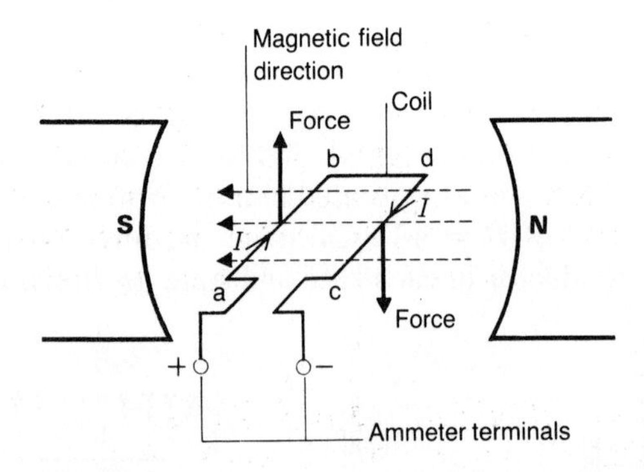

(c) Simplified diagram showing force on coil

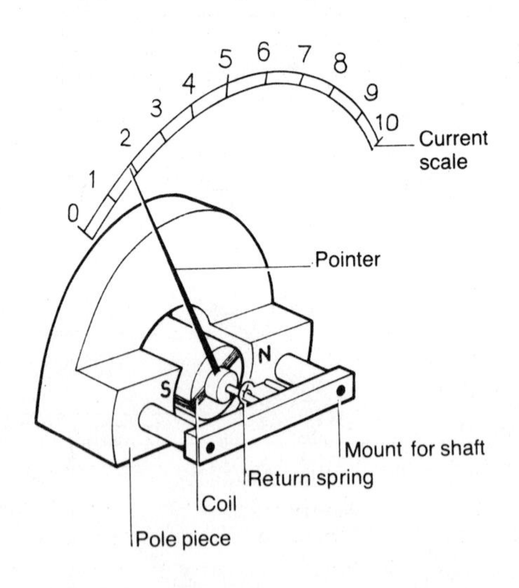

(d) Diagram of a practical instrument

Figure 14.13 Moving-coil meters

current, i.e. $F = BIL$, and this property provides a convenient means of measuring current (and voltage) which is applied practically in moving-coil meters.

The main features of a moving-coil ammeter are shown in Figure 14.13. The small coil through which the current to be measured flows is usually wound on a light aluminium former. The coil consists of copper wire supplemented with constantan to minimize any increase in resistance which could occur with temperature. Attached to the coil is a shaft which is mounted on steel pivots. The coil is free to rotate in the air gap between the curved pole pieces of a powerful permanent magnet and a ferromagnetic core. The magnet and core produce a radial form of magnetic field ensuring that the flux density B is at right angles to the coil current as the coil rotates, see (b).

When a current flows through the coil the conductor sections ab and cd, see (c), experience force. The force on ab is upwards and the force on cd is downwards. There are no forces on bd and ac as the current in these sections is parallel to the magnetic flux lines. The forces on ab and cd produce a turning moment which will cause the coil to rotate but this rotation is counterbalanced by a restoring moment in the opposite direction produced by a spring (sometimes two springs) attached to the shaft. In equilibrium, the moment produced by the current-carrying coil in the magnetic field produced by the magnet is equal to the restoring moment produced by the spring(s). The rotation of the coil is recorded by a pointer attached to the shaft moving over a linear scale which is calibrated in terms of current.

For current-measurement applications the terminals of the instrument are shunted with low-value resistances known as shunts. Typical shunt values are of the order of 0.01 Ω to 0.1 Ω for currents in the range 0 to 10 A, and 1 Ω to 0 Ω for currents in the milliampere range. The moving-coil meter can also be adapted for voltage measurement by inserting a high-value resistor in series with the meter. Typical values for this series resistor are 200 kΩ to 2 MΩ for voltage measurement in the range 0 to 100 V, and of the order of 20 kΩ for voltages in the 0 to 1 V range.

14.6 Operation of a simple dc electric motor

The force on a current-carrying conductor in a magnetic field forms the means by which electric energy can be converted to mechanical energy and provides the basis for the design of electric motors.

Figure 14.14(a) shows a diagram of a simple dc electric motor consisting of a coil abdc situated between the poles of a magnet. Attached to the coil is a shaft and the coil plus shaft are free to rotate. Current is supplied to the coil from an external dc source by means of a commutator, see (b). This consists essentially of a split conducting cylinder, the two segments of which are insulated from each other but are permanently connected to the opposite ends of the coil. Current is supplied to these segments and hence the coil by means of carbon graphite brushes as the coil rotates. The important function of the commutator is to reverse the direction of current flow and hence the direction of the forces on the coil conductors ab and cd each time the coil passes through the vertical.

When current is fed into the coil in the direction shown in Figure 14.14(c) we find on applying Fleming's left-hand rule that conductor ab experiences an upward force and conductor cd a downward force. This causes the coil to rotate. When the coil reaches the vertical position, the current in these two conductors is reversed by means of the commutator and hence the forces are also reversed in direction. Thus, the coil continues to rotate. Actually, when the coil is in the vertical position, the forces on the coil conductors ab and cd do not cause rotation. However, the rotational energy carries the coil over the vertical position and as the current is also reversed at this instant, rotation is maintained.

14.7 Electromagnetic induction: Faraday's and Lenz's laws

Electromagnetic induction is the phenomenon whereby an electromotive force can be generated in a circuit or conductor by causing the magnetic flux 'linking' the circuit or 'cutting' the conductor

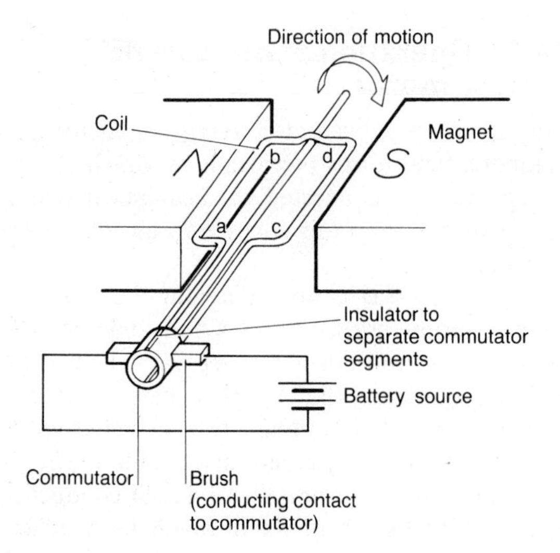

(a) *A simple dc motor*

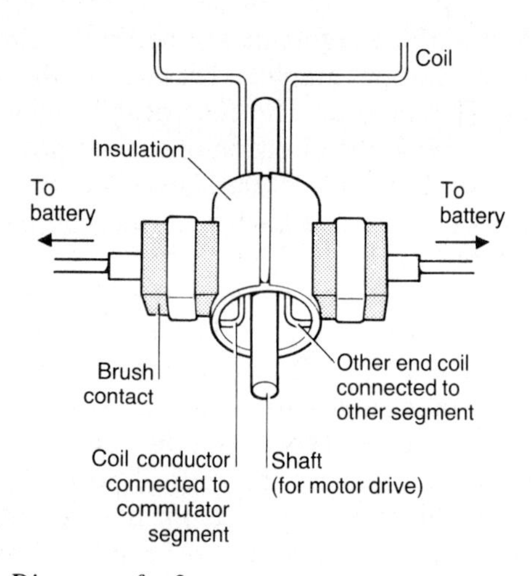

(b) *Diagram of a 2-segment commutator*

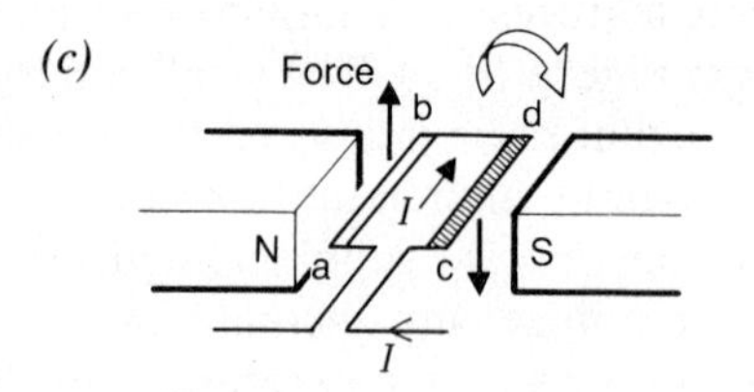

(i) Coil rotates due to forces on conductors ab and cd

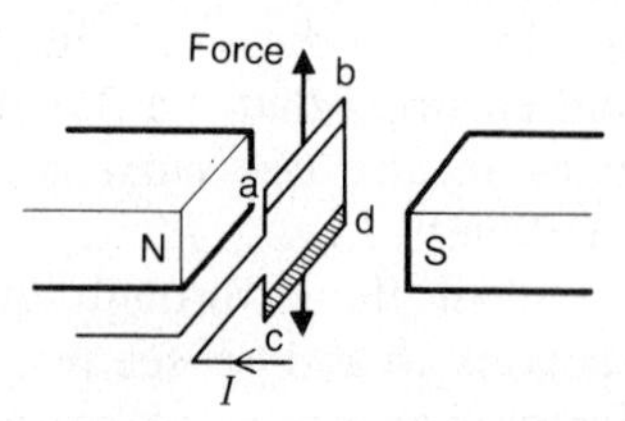

(ii) Coil goes through vertical dead and current direction reversed by commutator

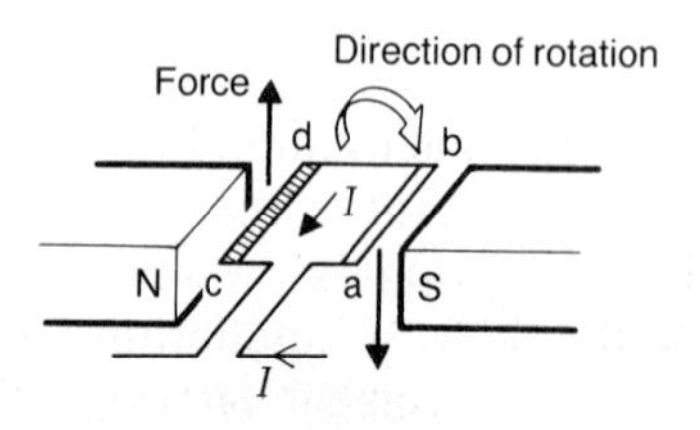

(iii) Coil one half revolution later than (i) shows reversed current direction in conductors ab and cd

Figure 14.14 Principle of a dc electric motor

to change in some way. The changing flux induces an e.m.f. Electromagnetic induction forms the basis whereby mechanical energy can be converted to electrical energy and is the fundamental means used to generate electricity in electrical power stations.

Figure 14.15 shows a simple experiment to illustrate the production of an e.m.f. and the resulting current flow by means of electromagnetic induction. An e.m.f. is induced in coil C by moving the bar magnet M in and out of the coil therefore causing a change in the magnetic flux produced by M linking the turns of the coil.

When the magnet is moved toward the coil the galvanometer G (a sensitive ammeter which in this case has a centre-zero scale) connected across the terminals of the coil gives a kick and registers a deflection. The deflection falls to zero as soon as the magnet comes to rest. If M is withdrawn from the coil, the galvanometer registers a deflection in the opposite direction. Although it may be difficult to observe experimentally, the magnitude of the galvanometer deflections depends on the speed of moving the magnet – the faster, the greater the deflection. By moving the magnet continuously in and out of the coil, th

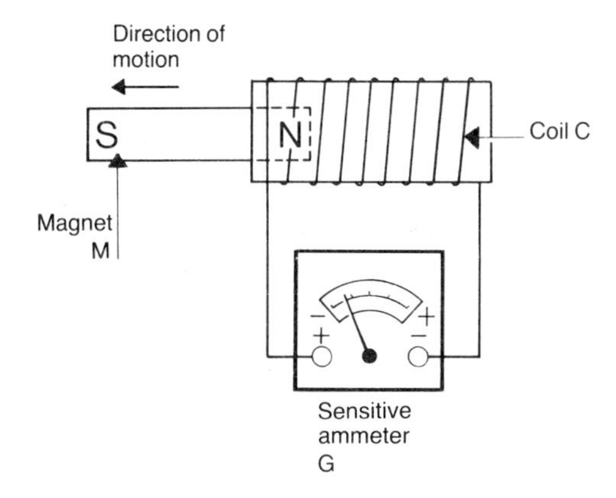

Figure 14.15 Experiment to demonstrate the e.m.f. in a coil due to changing magnetic flux linking the coil

galvanometer needle can be made to oscillate backwards and forwards about its zero position and a relatively large swing in needle deflection can be obtained. In fact, an identical experiment to this was undertaken in 1831 by Faraday in his classical work on electromagnetic induction.

From the observation of our experiment, we can make the following conclusions:

1. The movement of a magnet in a coil generates an e.m.f. and thus causes current to flow in the coil.
2. The direction of action of the e.m.f., and therefore the current induced in the coil, depends on the direction of motion of the magnet.

These observations form the basis of **Faraday's Law of electromagnetic induction**, which states:

The e.m.f. induced in a circuit is equal to the rate of change of magnetic flux linking the circuit or equivalently the rate at which the conductors in the circuit cut the magnetic flux.

In the case of a magnet moving in and out of a coil, the coil experiences a changing magnetic flux due to the magnet's motion, and the rate of change of flux 'linking' the coil produces an e.m.f. in the coil. If the magnet is held stationary and the coil is moved in its magnetic field, an e.m.f. is also induced in the coil. In both cases the coil experiences a changing magnetic flux and it is this that induces the e.m.f.

The direction of the induced e.m.f. is given by Lenz's Law, which states:

The direction of the induced e.m.f. is such as to cause a current, if the circuit is complete, in such a direction so as to set up magnetic flux in the opposite direction to the magnetic flux producing the e.m.f.; or equivalently, to set up a magnetic field in the circuit to oppose the motion of the source of magnetic field inducing the e.m.f.

The application of Lenz's law is illustrated in Figure 14.16. In (a) when the magnet is approaching the coil an e.m.f. and therefore current will be induced in the coil so as to produce a N pole at the coil end closest to the approaching N pole of the magnet thus opposing the magnet's motion since like poles repel. Work is done in moving the magnet and electrical energy is generated in the coil. When the magnet is withdrawn, the induced e.m.f. and current direction will be reversed to set up a S pole in the coil end and thus again oppose the motion of the magnet, as shown in (b).

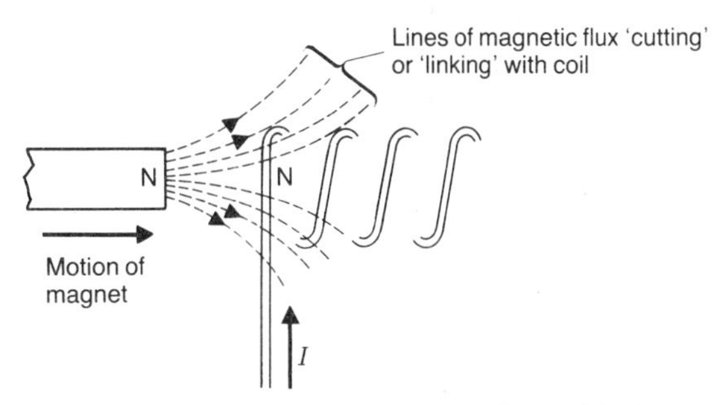

Direction of induced current to produce an effective 'N' pole in coil to oppose motion of magnet

(a)

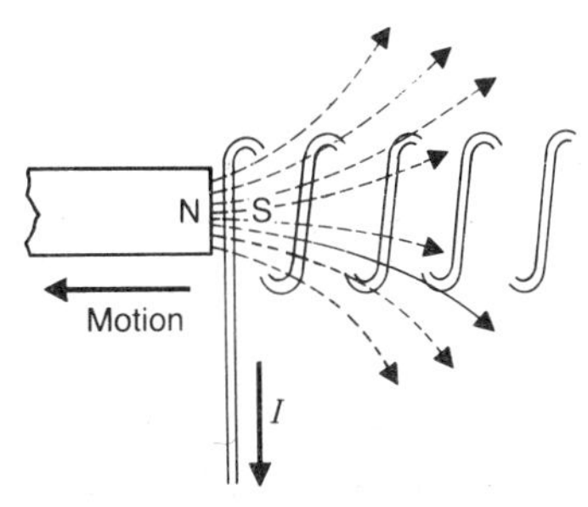

Direction of induced current to produce an effective 'S' pole in coil to oppose motion of magnet

(b)

Figure 14.16 Application of Lenz's law to determine the direction of induced e.m.f. and current flow

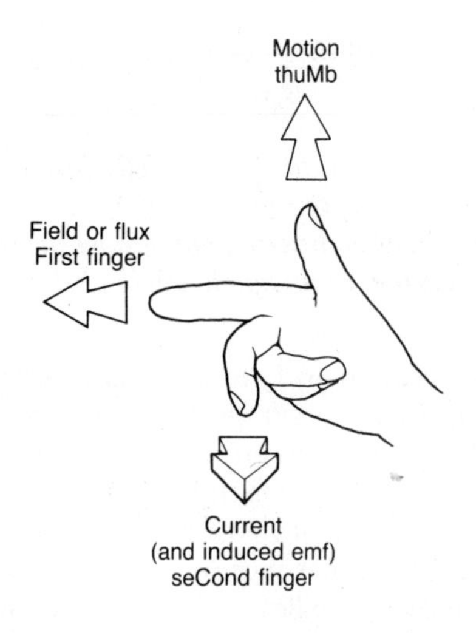

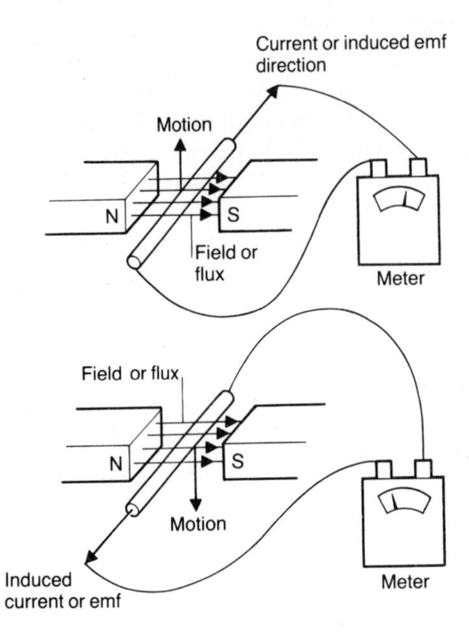

Figure 14.17 Fleming's right-hand rule

A useful rule for determining the direction of the induced e.m.f. when a conductor is moved through a magnetic field and cutting lines of magnetic flux is given by Fleming's right-hand rule, see Figure 14.17: Hold your right hand with the thumb and first two fingers at right-angles; point the first finger in the direction of the lines of magnetic flux, the thumb in the direction of motion, then the second finger will point in the direction of the induced e.m.f. and positive current, if this flows. Check by applying the rule, that you agree with the direction of the induced e.m.f. and current shown in the two examples of Figure 14.17(b).

Note Fleming's right-hand rule is used to determine *the direction of the induced e.m.f.* when a conductor *is moved* in a magnetic field. Fleming's left-hand rule, considered in section 14.4 is used to determine *the direction of the force* on a current-carrying conductor in a magnetic field.

14.8 Application of electromagnetic induction to generation of electricity: operation of a simple ac generator

Figure 14.18 shows a diagram of a simple ac generator in which an alternating e.m.f. is generated by rotating a coil in the magnetic field produced between the poles of a magnet. In such a generator, external mechanical energy is supplied to rotate the coil and this energy is converted to electrical energy by means of electromagnetic induction. Today, the generation of virtually all mains electrical power utilizes this principle.

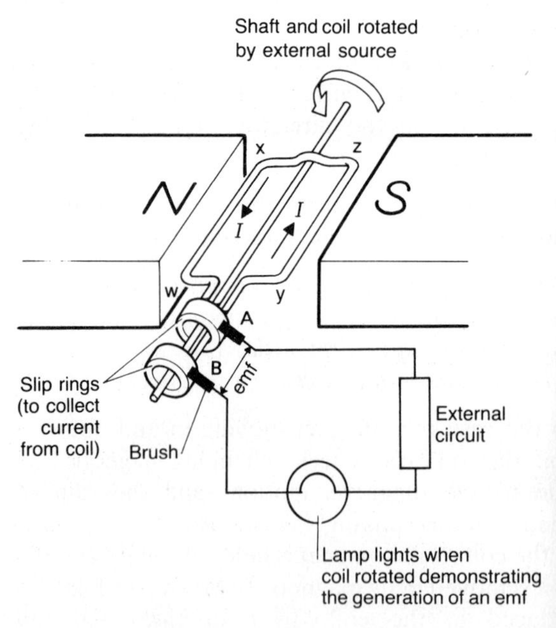

Figure 14.18 Sketch of simple ac generator. The ac e.m.f. induced in the coil wxyz as it rotates is developed across the slip-rings A and B. The coil conductor from side w is connected to A and the conductor from side y is connected to B

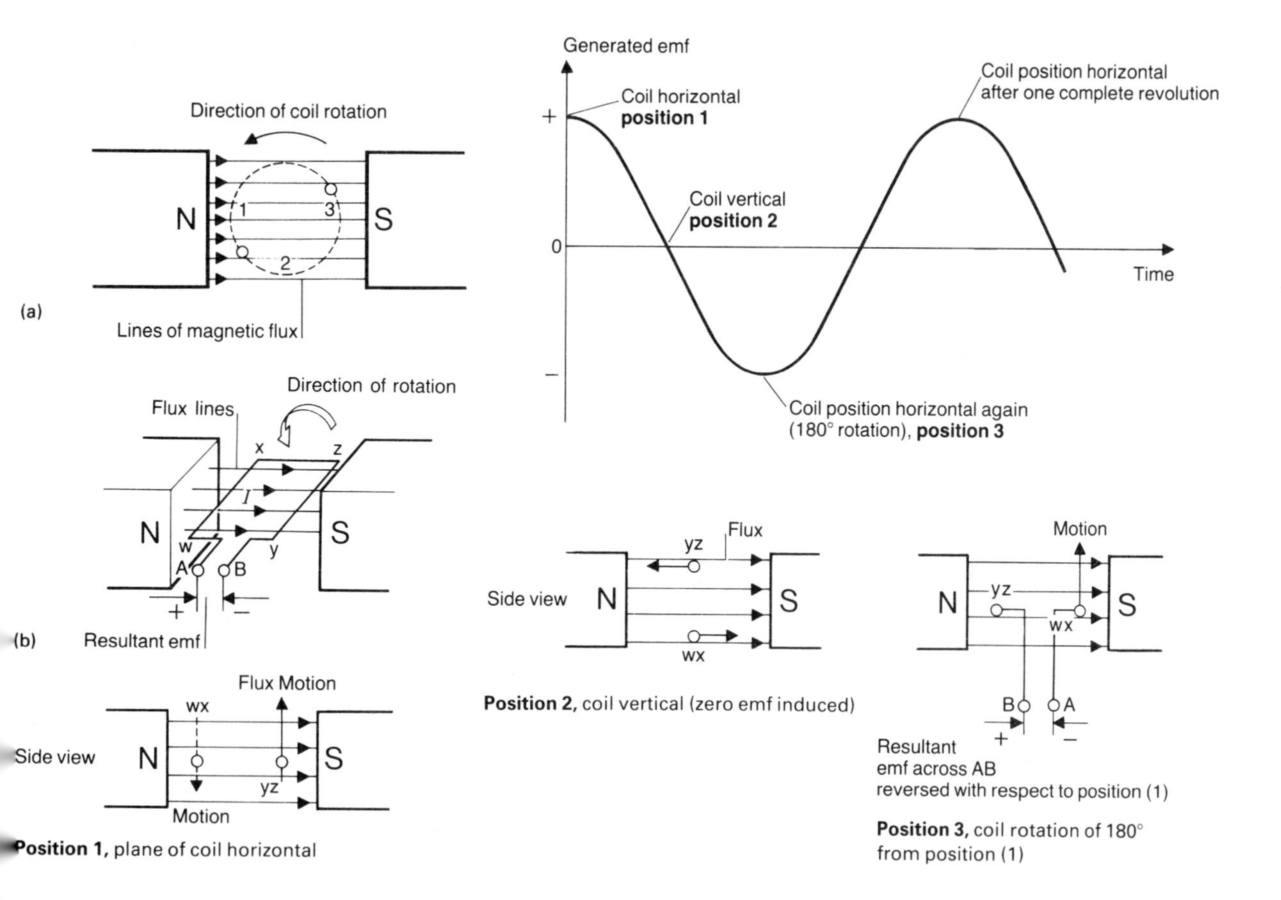

Figure 14.19 Action of ac generator
(a) Cross-sectional view of ac generator and variation of e.m.f. it produces with time and rotational position
(b) An alternating e.m.f. is induced in the coil as it cuts magnetic flux; the direction of the e.m.f. in conductors wx and yz may be determined using Fleming's right-hand rule. Resultant e.m.f. equals the sum of the e.m.f.s induced in wx and yz and is developed across terminals A-B (these normally being connected to slip-rings). Three consecutive coil positions are shown

The action of this simple ac generator may be explained with reference to Figure 14.19, where (a) shows the variation of e.m.f. generated as the coil is rotated. The e.m.f. is induced in the conductors wx and yz as they cut the lines of magnetic flux. Maximum e.m.f. is generated when the coil is in the horizontal position since in this position wx and yz are cutting the flux at the maximum rate. When the coil is passing through the vertical no e.m.f. is generated since at this instant wx and yz are moving instantaneously parallel to the flux lines and not cutting flux. The directions of the induced e.m.f.s in wx and yz can be determined by applying Fleming's right-hand rule. Note that the e.m.f.s induced in the two conductors reinforce each other. Three consecutive positions in the rotation cycle are shown in more detail in Figure 14.19(b):

1 In position 1 when the coil is horizontal, the e.m.f. generated in both conductors is a maximum, the conductors are cutting flux at the maximum rate and the e.m.f.s in wx and yz aid each other making terminal A positive with respect to B.
2 In position 2 when the coil is vertical, zero e.m.f. is generated since both conductors are moving parallel to the lines of magnetic flux and are therefore cutting no flux. Zero e.m.f. is generated across A–B. Between positions 1 and 2 the e.m.f. falls from maximum to zero.
3 On moving through the vertical position the

Answer block:

Question no.	1				2		
	(a)	(b)	(c)	(d)	(a)	(b)	(c)
Answer							
Question 3	Answer Qu 3 ↓						
Question 4	Answers Qu 4						
	(a)		(b)		(c)		
	(d)						

direction of the e.m.f.s induced in wx and yz are both reversed, so A now becomes negative with respect to B. The magnitudes of the e.m.f. increase as the rate of cutting flux increases and become maximum in value when the coil passes through the horizontal position.

Thus the e.m.f. of the generator varies cyclically, with maximum amplitude being generated whenever the coil passes through the horizontal, and zero when the coil passes through the vertical with the direction of the e.m.f. also changing at these instants. The frequency of the ac voltage produced is, of course, equal to the number of rotations the coils makes per second, e.g. if the coil is rotated at fifty revolutions per second, the ac frequency is 50 Hz.

Test 14

This test may be used as a basic self-assessment test to check whether you have understood the main facts of the final Chapter 14 on **Electromagnetic effects and applications**. All answers are to be entered in the answer block.

Qu. 1 Enter a tick (√) in the answer block if yo consider the statement is correct; enter cross (×) if you consider the statement in any way incorrect.

(a) A conductor carrying current gene ates its own magnetic field and wi experience a force if placed in magnetic field.

(b) Ferromagnetic materials exhib strong magnetic effects and incluc iron and copper.

(c) The e.m.f. induced in a coil conductor is proportional to the ra of change of magnetic flux throug the coil or equivalently the rate which the conductor cuts magnet flux.

(d) Electromagnetic induction is appli in electric generators which essenti ly act to convert mechanical ener to electrical energy by rotating a c in a magnetic field.

Qu. 2 The formula for the force on a conduct carrying a current, when placed in magnetic field is given by

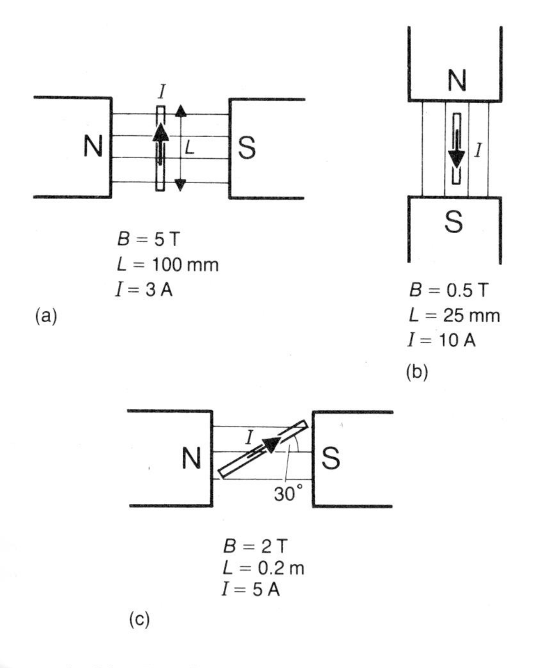

ure 14.20 For Qu 2

$$F = BIL \sin\theta$$

where θ = angle between the directions of magnetic flux density and current flows.

Calculate the magnitude of the force for the cases shown in Figure 14.20(a), (b) and (c).

. 3 In Figure 14.21 the magnet is moved towards the solenoid. Explain concisely what happens in the solenoid circuit.

. 4 Figure 14.22(a) shows the diagram of a simple ac generator and (b) the cross-sectional view of six successive stages in the coil rotation, starting in position 1 with the coil passing through the vertical. The coil is rotated at 100 revolutions per second and the e.m.f. generated across A–B at the instant of position 2, is 25 volts.

(a) In which of the six positions is the instantaneous value of the generated e.m.f. zero.

(b) Determine the value of the e.m.f. generated when the coil passes through position 4.

(c) What is the frequency of the e.m.f. generated?

(d) Sketch the waveform of the e.m.f. generated, starting from position 1 up to position 6.

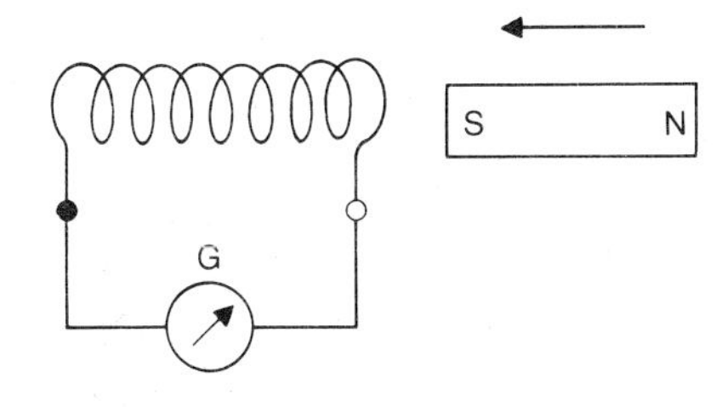

re 14.21 For Qu 3

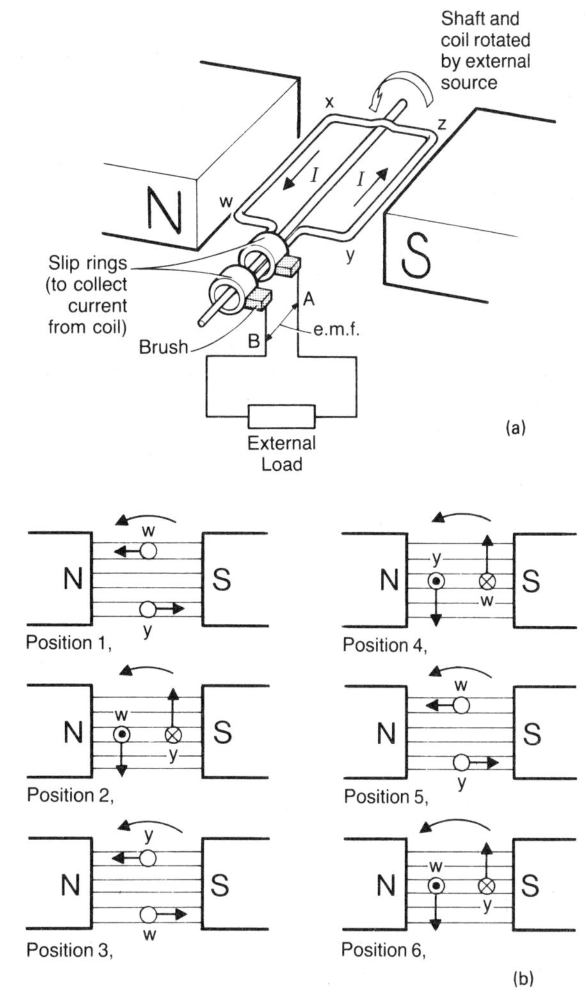

Figure 14.22 For Qu 4
(a) AC generator
(b) Cross-sectional view of six consecutive coil positions

Problems 14

1 Sketch the magnetic field produced by (a) a bar magnet, (b) a straight conductor carrying current and (c) a solenoid carrying current. Describe how the field could be plotted for the case of a bar magnet.

2 State the factors on which the force on a current-carrying conductor in a magnetic field depend and describe a simple experiment to demonstrate the magnitude and direction of such a force.

3 State the formula for the force on a current-carrying conductor in a magnetic field. Define the meaning of all terms and state their units.
A straight wire of length 200 mm and carrying a current of 15 A is at right angles to a magnetic field of 0.4 T. Calculate the force on the wire.

4 Calculate the magnitude and state the direction of the force for the current-carrying conductors shown in Figure 14.23. In each case the conductor length is 50 mm, the current is 4 A and the magnetic flux density is 2 T.

5 (a) A straight wire of length 150 mm and carrying a current of 5 A is at right-angles to a uniform magnetic field. If the force on the wire is 0.2 N calculate the magnetic flux density of the field.

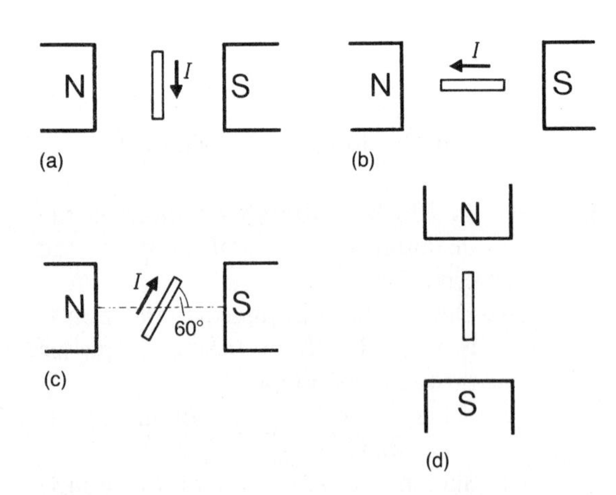

Figure 14.23 For problem 4

(b) A straight wire of length 250 mm is at an angle of 60° to a uniform magnetic field of flux density 3 mT. If the force on the wire due to the magnetic field is 20 mN calculate the current flowing in the conductor.

6 Draw a labelled sketch showing the essential features of a moving-coil ammeter and explain the action of the meter.

7 Figure 14.24 shows a single-turn rectangula coil situated between the poles of a magne which produces a uniform magnetic field o flux density B = 0.5 T. Sides ab and cd of th coil are of 20 mm length and sides bd and a 15 mm. If the coil carries a dc current of I = 5 A, calculate the forces on each of the coi sides, the turning moment produced and th direction the coil would rotate for the case o I in the direction abdc.

8 (a) State what is meant by electromagneti induction and describe a simple experi ment to illustrate it.
(b) State Faraday's law of electromagneti induction.
(c) State Lenz's law and describe how it ca be applied to determine the direction c the e.m.f. induced in a coil when the flu linking the coil is changing.

9 Draw a labelled diagram illustrating th essential features of a simple ac generato Explain its basic mode of operation an sketch a graph of e.m.f. generated vers time as the coil rotates.

10 The e.m.f. produced by a simple ac gene ator is given by

$v = NAB\omega\cos\theta$ volts

where B = magnetic flux density, T
N = number of turns of coil
A = coil area, m^2
$\theta = \omega t$ radians, t = time
ω = angular speed at which t coil is rotated, rad

A coil of 500 turns and cross-sectional ar 900 mm^2 is rotated at a constant angu speed of $2\pi \times 50$ rad/s (50 revolutions p second) between the poles of a magnet whi

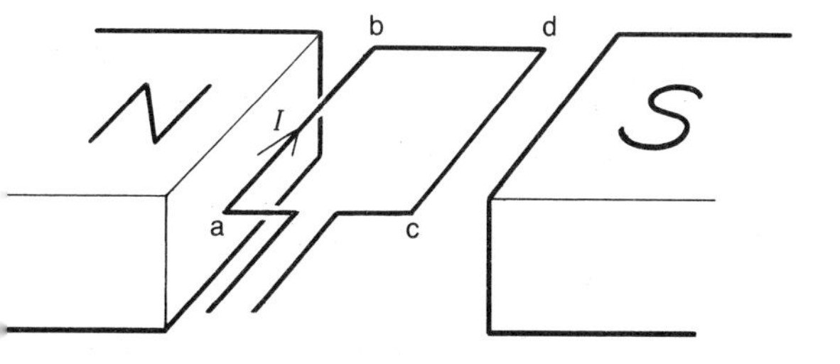

ure 14.24 *For problem 7*

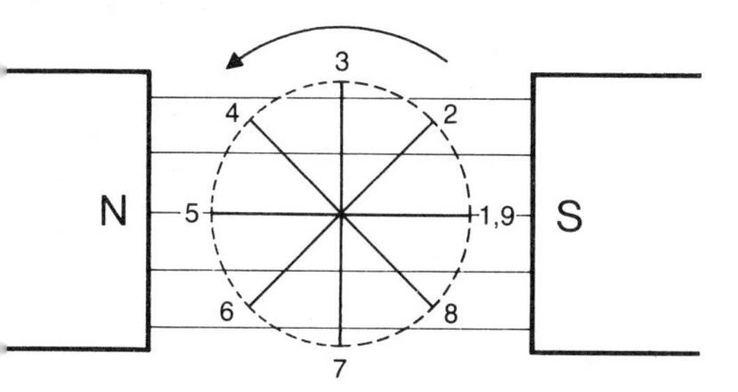

ure 14.25 *For problem 10*

produces a field of flux density 0.2 T. Figure 14.25 shows nine consecutive positions spaced 45° apart in the rotation cycle, starting at 1 with the coil horizontal.

Describe the action of the generator as it passes through these positions and calculate the value of the e.m.f. induced across the coil terminals for each position. Plot a curve of e.m.f. versus time for $t = 0$ to $t = 40$ ms. Calculate also the peak amplitude, frequency and periodic time of the e.m.f. waveform.

Answers to Tests and Problems

Chapter 1 Oxidation

Test 1

Question no.	1				2			3			
	(a)	(b)	(c)	(d)	(a)	(b)	(c)	(a)	(b)	(c)	(d)
Answer	√	×	√	√	√	√	√	×	√	√	√
Question no.	4							5			
	(a)	(b)	(c)	(d)	(e)	(f)	(g)	(a)	(b)	(c)	(d)
Answer	√	×	×	√	√	√	√	√	√	√	√

Problems 1

2 (a) copper and oxygen
 (b) oxidation, copper oxide

3 266.7 g of oxygen, 366.7 g of carbon dioxide

4 (a) will not rust; (b) will rust; (c) will rust

5 oxygen and water

Chapter 2 Elasticity and Hooke's law

Test 2

Question no.	1					2
	(a)	(b)	(c)	(d)	(e)	
Answer	√	√	√	×	√	(a) compressive (b) shear (c) ductile (d) extension (strain) (e) constant Young's modulus

Question no.	3				4		
	(a)	(b)	(c)	(d)	(a)	(b)	(c)
Answer	0.6 mm	150 N	0.4 mm	0.25 mm	17 mm	84 N	120 N

Problems 2

2 (a) $44\ N/mm^2$; (b) 3.182 mm; (c) 165 N
3 (b) (i) 0.533 mm; (ii) $1875\ N/mm^2$; ($1.875\ GN/m^2$)
4 $50\ MN/m^2$; $125\ mm^2$

Chapter 3 Forces, moments and static equilibrium

Test 3

Question no.	1						2	3	4
	(a)	(b)	(c)	(d)	(e)	(f)			
Answer	√	√	√	×	√	√	23.3 N	58 N ∠23°	3 m 640 N 64 kg

Problems 3

3 (a) 250 N; (b) 750 N, 5 m from P
4 3250 N, 6250 N
5 $T_1 = 400$ N, $T_2 = 0$N
6 112 N ∠1.25°
7 125 N ∠37°
8 156 N ∠67.5°
9 7.1 N

Chapter 4 Pressure in fluids

Test 4

Question no.	1					2	3		
	(a)	(b)	(c)	(d)	(e)		(a)	(b)	(c)
Answer	√	×	×	×	√	1.1068×10^6 N/m^2 (Pa)	1.173×10^5	0.96×10^5	1.013×10^5

Note:

Qu. 2 Pressure = 1.005525×10^6 + atmospheric pressure (1.013×10^5)
= 1.1068×10^6 Pa ≈ 10.9 × atmospheric pressure.

Qu. 3 (a) 16×10^3 Pa above atmospheric pressure = 1.173×10^5 Pa
(b) 5.3×10^3 Pa below atmospheric pressure = 0.96×10^5 Pa
(c) equal to atmospheric pressure = 1.013×10^5 Pa

Problems 4

1 (a) 3.3109×10^4 N, $1.1036 \times 10^4\ N/m^2$ (or Pa)
(b) $0.5739 \times 10^4\ N/m^2$ or $5.739\ kN/m^2$
4 0.998×10^5 Pa
5 6.905×10^3
6 (a) 99.56°C; (b) 96.86°C

Chapter 5 Speed, velocity and distance–time graphs

Test 5

Question no.	1					2				3		
	(a)	(b)	(c)	(d)	(e)	(a)	(b)	(c)	(d)	(a)	(b)	(c)
Answer	√	√	√	√	×	2.273 m/s	6.67 m/s	40.8 km	500 s	4.83 m/s	24 m	4.3 m/s

Note: In 1(e) the speed in orbit is approximately constant but the velocity is continually changing direction, hence although the magnitude of the velocity is constant its direction is not.

Problems 5

1 (a) 20 m/s; (b) 5.95 m/s
2 (a) 1.51 s; (b) 1.28 s
3 (a) 0.75 m/s; (b) (i) 0.63 m/s, (ii) 16.5 s

Chapter 6 Acceleration, velocity–time graphs and solving problems in linear motion

Test 6

Question no.	1					2				3			4		
	(a)	(b)	(c)	(d)	(e)	(a)	(b)	(c)	(d)	(a)	(b)	(c)	(a)	(b)	(c)
Answer	√	√	√	√	×	20 m	24 m	−2 m/s^2	7.33 m/s	30 m/s	45 m	6.32 s	4 m/s^2	24 m/s	128 m

Problems 6

1 (b) 20 kg; (c) $-2\ m/s^2$, 400 m
2 (a) $1.67\ m/s^2$, 0, $-2.5\ m/s^2$; (b) 3750 m, 37.5 m/s
3 (a) $7.7\ m/s^2$; (b) $-40\ m/s^2$; (c) 6200 m, 124 m/s
4 (a) 37.3 m/s; (b) 70.8 m
5 15 m/s; 45 m
6 $4\ m/s^2$; 125 m; 350 m
7 300 m/s; 4500 m
8 (a) 50 N; (b) 25 m; (c) 5 m
10 200 N

Chapter 7 Waves and wave motion

Test 7

Question no.	1				2		3	
	(a)	(b)	(c)	(d)	(a)	(b)	(a)	(b)
Answer	√	√	√	×	300 m	12 GHz	3.03 s	5000 m/s

Problems 7

2 331.3 m/s
3 lf: 10 000–1000 m; mf: 1000–100 m; hf: 100–10 m; vhf: 10–1 m; uhf: 1–0.1 m; microwaves: 0.3 m–1 mm.
4 29 mm; (i) 6.9 ms; (ii) 345 ms

Chapter 8 Work, energy and power

Test 8

Question no.	1					2			3		4		
	(a)	(b)	(c)	(d)	(e)	(a)	(b)	(c)	(a)	(b)	(a)	(b)	(c)
Answer	√	×	√	√	√	200 J	350 J	500 J	0.7 (70%)	50 kJ	80 W	48 kJ	12 kJ

Problems 8

2 (a) 125 kJ; (b) 7420 J; (c) 3464 J.
3 (a) 210 W; (b) 0.9 or 90%; (c) 181.8 kJ.
5 3450 J
6 (a) 1000 J; (b) 500 J

Chapter 9 Heat and temperature-change effects

Test 9

Question no.	1				2	3	4	
	(a)	(b)	(c)	(d)			(a)	(b)
Answer	√	√	×	×	1365 kJ	23.5°C	2.4 mm	83.3°C

Problems 9

2 84.764 MJ; 7.85 hours
3 453.5 $J\,kg^{-1}\,K^{-1}$
4 2 MJ
5 (a) 200 kJ; (b) 3.34 MJ; (c) 4.18 MJ; (d) 22.6 MJ; 2.81 hours or 10 107 s
6 18.63 kg
8 72 kN
9 $20 \times 10^{-6}\,°C^{-1}$

Chapter 10 Introduction to electricity: electrical current and voltage

Test 10

Question no.	1					2		
	(a)	(b)	(c)	(d)	(e)	(a)	(b)	(c)
Answer	√	√	√	√	√	340 V	10 ms	50 Hz

		Answers
Question 3	(a)	Battery, solar cell
	(b)	Alternator (Electromagnetic ac generator)
	(c)	Electroplating, powering electronic equipment
	(d)	Ease of distribution and generation

Problems 10

3 (a) 2 A, 1000 Hz; (b) 5 V, 125 kHz; (c) 10 V, 10 Hz; (d) 10 mA, 50 Hz.

4 (a) 7 V; (b) 50 V

Chapter 11 Resistance and series and parallel resistive circuits

Test 11

Question	1				2		3		
	(a)	(b)	(c)	(d)	(a)	(b)	(a)	(b)	(c)
Answer	√	×	×	×	0.3 A	6 V	20 Ω	20 V	40 Ω

Question	4			5			6	
	(a)	(b)	(c)	(a)	(b)	(c)	(a)	(b)
Answer	7.5 A	2.5 A	10 A	15 kΩ	80 Ω	60 Ω	0.002 $°C^{-1}$	110 Ω

Problems 11

1 (a) 2 Ω; (b) 200 Ω; (c) 400 kΩ
3 0.02 A or 20 mA; 12 V
4 (a) 2.5 mA; (b) 25 V, 75 V
5 150 Ω, 0.33 A
6 100 Ω, 0.1 A
7 (a) 200 V; (b) 20 V; (c) 0.67 A, 3.33 A;
(d) 220 V
8 (a) 140 Ω; (b) 0.5 A; (c) 50 V; (d) 200 mA, 250 mA
9 (a) 40 V; (b) 36 V
11 (a) 60 Ω, 80.08 Ω; (b) 0.007$°C^{-1}$
12 (a) 191 mA; (b) 198 mA

Chapter 12 Power in electrical circuits

Test 12

Question no.	1			2			3		
	(a)	(b)	(c)	(a)	(b)	(c)	(a)	(b)	(c)
Answer	√	√	√	0.25 A	0.5 A	8.3 A	62.5 Ω	18 MJ	50p

Problems 12

1 (a) 0.08 A; (b) 0.384 W; (c) 0.64 W
2 (a) 25 W, 12.5 W; (b) 37.5 W
3 (a) 30 Ω, 1 A; (b) 13.33 W
4 (a) 1 A; (b) 3 A; (c) 15 A

Chapter 13 Electrochemical effects and applications

Test 13

Question no.		Ans	Question no.		Answer Qu. 2
1	(a)	√	2	(a)	Anode
	(b)	√		(b)	Cathode
	(c)	√		(c)	Nickel ions (Ni^{++}) flow to cathode Sulphate ions (SO_4^{--}) flow to anode
	(d)	√			
	(e)	√		(d)	Nickel metal is deposited
3	(a)	10 V		(e)	Sulphate ions react with nickel electrode and nickel ions pass back into solution
	(b)	8.6 V			
	(c)	0.4 Ω		(f)	Electrolysis

Problems 13

3 4.22 A
5 5 V, 0.4 Ω (approx)
6 1.5 V, 2 Ω
7 (a) 10 V; (b) 9.6 V
8 (a) 2.89 A (b) 1.04 A
9 (a) 11.6 Ω (b) 1.36 Ω

Chapter 14 Electromagnetic effects and applications

Test 14

<table>
<tr><td rowspan="2">Question no.</td><td colspan="4">1</td><td colspan="3">2</td></tr>
<tr><td>(a)</td><td>(b)</td><td>(c)</td><td>(d)</td><td>(a)</td><td>(b)</td><td>(c)</td></tr>
<tr><td>Answer</td><td>√</td><td>×</td><td>√</td><td>√</td><td>1.5 N</td><td>0</td><td>1 N</td></tr>
<tr><td>Question 3</td><td colspan="7">Answer Qu. 3</td></tr>
<tr><td></td><td colspan="7">An e.m.f. is induced in the solenoid, current flows and the meter deflects while the magnet is in motion</td></tr>
<tr><td>Question 4</td><td colspan="7">Answers Qu. 4</td></tr>
<tr><td rowspan="2"></td><td colspan="7">(a) 1,3,5; (b) −25 V; (c) 100 Hz</td></tr>
<tr><td colspan="7">(d) 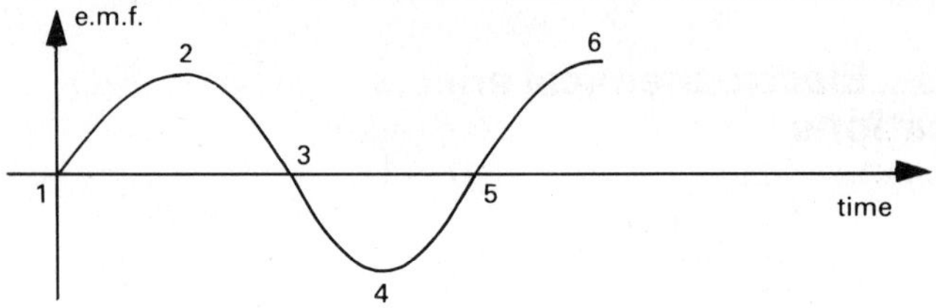</td></tr>
</table>

Problems 14

3 1.2 N

4 (a) 0.4 N, out of paper; (b) 0;
(c) 0.346 N, into paper; (d) 0

5 (a) 0.267 T; (b) 30.8 A

7 0.05 N, 7.5×10^{-4} Nm anti-clockwise

10 28.3 V, 50 Hz, 20 ms

Index